Hernán Javier Guardia Morán

General Economic Geography

Hernán Javier Guardia Morán

General Economic Geography

An analysis from practice

ScienciaScripts

Imprint

Any brand names and product names mentioned in this book are subject to trademark, brand or patent protection and are trademarks or registered trademarks of their respective holders. The use of brand names, product names, common names, trade names, product descriptions etc. even without a particular marking in this work is in no way to be construed to mean that such names may be regarded as unrestricted in respect of trademark and brand protection legislation and could thus be used by anyone.

Cover image: www.ingimage.com

This book is a translation from the original published under ISBN 978-613-9-40287-8.

Publisher:
Sciencia Scripts
is a trademark of
Dodo Books Indian Ocean Ltd. and OmniScriptum S.R.L publishing group

120 High Road, East Finchley, London, N2 9ED, United Kingdom
Str. Armeneasca 28/1, office 1, Chisinau MD-2012, Republic of Moldova, Europe
Printed at: see last page
ISBN: 978-620-7-66174-9

HUMAN GEOGRAPHY
(GENERAL ECONOMIC GEOGRAPHY AND GEOGRAPHY
OF PANAMA)

DRAFTED BY:
HERNÁN JAVIER GUARDIA MORÁN

FOREWORD

Economic geography is a fascinating field of study that invites us to explore the complex interaction between geographic space and human economic activities. In this book, we immerse ourselves in a journey through the various dimensions of this discipline, from classical theories to contemporary trends that shape our globalized world.

In today's era, where borders are blurring and distances are shortening thanks to technology, understanding how geography influences the distribution of resources, the production of goods and services, international trade and regional development becomes crucial to understanding the challenges and opportunities facing the global economy.

Throughout these pages, we will discover how factors such as climate, topography, natural resources, infrastructure and geographic location impact the competitiveness of nations, productive specialization, labor mobility and people's quality of life.

From Alfred Weber's classical theories of industrial location to contemporary models of business clusters and global value chains, we will explore the different currents of thought that have shaped economic geography and contributed to our understanding of the ever-evolving economic world.

This book not only seeks to provide an in-depth theoretical analysis, but also to connect these concepts with concrete examples and case studies that illustrate the practical relevance of economic geography in areas such as sustainable development, urban planning, natural resource management, and public policy formulation.

As we embark on this intellectual journey, I invite the reader to question, reflect and discover how geography and economics intertwine their paths to shape the world we live in and challenge us to imagine a more equitable, sustainable and prosperous future for all. Dr. Ramiro Campos. University Professor

Table of Contents

INTRODUCTION

Discover the world through Economic Geography!

Embark on a fascinating journey through the world of Economic Geography, where the interaction between human activity and space becomes the stage for our development. Explore how the distribution of resources, territorial organization, and physical and human factors shape the production, distribution and consumption of goods and services on a global scale.

In this brochure you will find:

The fundamentals of Economic Geography: Decipher the key concepts that will allow you to understand the relationship between the economy and space.

Natural Resources: The Earth's Treasure: Discover the importance of natural resources as the basis of economic activity and their role in sustainable development.

Energy: The Engine of Progress: Explores the fundamental role of energy in the modern economy and the challenges facing its supply and consumption.

The Primary Sector: The Basis of Life: Delve into the world of agriculture, livestock, fisheries, forestry and mining, and understand their vital role in the economy and food security.

Globalization and International Economics: Analyzes the trends that connect the world's economies and how Economic Geography contributes to understanding them.

Case Studies: Real World Examples - Immerse yourself in concrete examples that illustrate how Economic Geography concepts apply in real life.

Economic Geography emerges as a discipline that analyzes the relationship between economic activity and space. It studies how the distribution of resources, territorial organization and physical and human factors influence the production, distribution and consumption of goods and services.

The importance of economic geography also has a practical application in the following fields: urban and regional planning: optimizing the location of industries, infrastructure and services for a balanced development, natural resource management, i.e. sustainable use of resources and mitigation of environmental impact, promotion of international trade to identify competitive advantages and opportunities in the global market, reduction of poverty and inequalities: implementing strategies for local and regional development.

Natural resources as the basis of economic activity:

Natural resources are elements found in nature that can be used to satisfy human needs. They are classified into renewable (such as solar energy and wood) and non-renewable (such as oil and minerals).

Energy is the ability to do work. It is an essential element in the functioning of modern society, powering industry, transportation, agriculture and everyday life.

The primary sector of the economy comprises economic activities that obtain resources directly from nature. It includes agriculture, livestock, fishing, forestry and mining.

CHAPTER 1: Economic Geography and its Practical Importance

1.1. Definition and content of Economic Geography

1.1.1. Definition of Economic Geography:

Economic geography is a discipline that studies the relationship between economic activity and space. It analyzes how the distribution of resources, territorial organization and physical and human factors influence the production, distribution and consumption of goods and services on a global scale.

In other words, economic geography seeks to understand how the economy is organized in space, how geographic characteristics condition economic activities, and how economic decisions impact the environment.

There are many other authors who have given meaning to this science, among them:

1. Samuelson, P. A. (1973). Economia. Madrid: McGraw-Hill.

Definition: "Economic Geography is the science that studies the relationship between economic activity and space".

2. Krugman, P. R., & Obstfeld, M. (2018). International economics: theory and policy. Barcelona: Pearson Education.

Definition: "Economic Geography analyzes how the distribution of natural resources, territorial organization, and physical and human factors influence the production, distribution, and consumption of goods and services on a global scale".

3. Morrill, R. L. (1981). Spatial dynamics of commodity production. Cambridge, MA: MIT Press.

Definition: "Economic Geography is a discipline that studies the spatial distribution of economic activities and how this distribution is affected by geographical and economic factors".

4. Clark, G. (1997). Why nations fail: The origins of power, prosperity, and poverty. New York: PublicAffairs.

Definition: "Economic Geography examines how geographic features, such as climate, topography, and location, influence the economic development of nations".

5. Krugman, P. R. (1998). Geography and trade. Cambridge, MA: MIT Press.

Definition: "Economic Geography analyzes how geographic factors, such as distance and transportation costs, affect international trade".

6. Henderson, V. L. (2001). Urban development: Economics, planning, and politics. New York: Oxford University Press.

Definition: "Economic Geography studies the location of economic activities in space, including the concentration of industries, urban centers and infrastructure".

7. Davis, H. K. (2005). World cities: A global network. Princeton, NJ: Princeton University Press.

Definition: "Economic Geography analyzes globalization and how it affects the spatial organization of economic activity, including the location of multinational corporations and foreign investment flows."

8. Duranton, G., & Puga, D. (2004). Urban agglomeration and economic development. Cambridge, MA: MIT Press.

Definition: "Economic Geography examines how urban agglomerations, such as large cities, can drive economic growth and innovation".

9. Fujita, M., Krugman, P. R., & Venables, A. J. (1999). The spatial economics of urbanization. Cambridge, MA: MIT Press.

Definition: "Economic Geography analyzes the location patterns of economic activities and how these patterns explain the formation and growth of cities".

10. Rodríguez-Pose, A. (2012). The spatial economics of development. Cheltenham, UK: Edward Elgar Publishing.

Definition: "Economic Geography examines how the spatial distribution of economic activity can contribute to economic development, especially in less favored regions".

We have seen how each author gives it a specialized or globalizing definition in which many social, natural and cultural elements come into play.

1.2. Content of Economic Geography:

The main topics addressed by Economic Geography are:

Distribution of natural resources: Analyzes the location and availability of natural resources, such as minerals, water, land and forests, and how these resources influence economic activities.

Territorial organization of economic activity: Studies the concentration and dispersion of economic activities in space, including the location of industries, urban centers, infrastructure and agricultural areas.

Globalization and international economics: Examines international trade flows, foreign investment, the location of multinational companies and the effects of globalization on the economy and territory.

Regional development and spatial inequalities: Analyzes the disparities in economic development between regions and countries, and how Economic Geography can contribute to the reduction of these inequalities.

Environmental sustainability and sustainable development:

Examines the relationship between economic activity and the environment, and how economic geography can contribute to more sustainable development.

a. Evaluation and main approaches to Economic Geography.

Economic Geography is a fundamental discipline for understanding the functioning of the economy in today's world. Its study makes it possible:

Identify economic opportunities: By analyzing the distribution of resources and territorial characteristics, potential areas for economic development can be identified.

Designing effective public policies: Understanding the geographic factors that influence the economy makes it possible to design more effective public policies to promote economic development and poverty reduction.

Mitigate environmental impacts: The analysis of the relationship between economic activity and the environment makes it possible to identify strategies to minimize negative impacts and promote sustainable development.

Understanding global economic trends: Economic Geography provides tools to analyze global economic trends and their effects on different regions and countries.

Throughout its history, Economic Geography has developed different approaches to the study of the relationship between economy and space. Some of the most important approaches include:

Environmental determinism: This approach holds that the physical characteristics of the environment largely determine economic activity.

Environmental Determinism:

Environmental determinism, also known as geographic determinism, is an approach that emerged in the 19th century and argues that the physical characteristics of the environment, such as climate, soil and topography, largely determine economic activity and social development.

Authors:

Ratzel, Friedrich: German geographer considered the father of environmental determinism.

Huntington, Ellsworth: American geographer who popularized the concept of "determining climates".

Analysis of the problem:

Environmental determinism has been criticized for its deterministic view and for not considering the human capacity to adapt and modify the environment. It is considered a simplistic approach that does not take into account the complexity of social, political and cultural factors that influence the economy.

Possibilism: This perspective states that the environment offers possibilities for economic activity, but it is human decisions that determine how these possibilities are exploited.

Possibilism:

Possibilism, in response to environmental determinism, emerged in the early 20th century and proposes that the environment offers possibilities for economic activity, but it is human decisions that determine how these possibilities are exploited.

Authors:

Blache, Vidal de la: French geographer who introduced the concept of "possibilism".

Febvre, Lucien: French geographer who emphasized the importance of "technologies" and "ways of life" in the relationship between man and the environment.

Analysis of the problem:

Possibilism has been recognized for its more flexible and dynamic approach, recognizing the active role of human action in the transformation of space. However, some critics point out that this approach may underestimate the influence of the environment and the constraints it imposes on economic activities.

Spatial economics: This approach uses mathematical and statistical tools to analyze the distribution of economic activity in space.

Spatial economics:

Spatial economics, developed in the mid-20th century, uses mathematical and statistical tools to analyze the distribution of economic activity across space. It focuses on understanding how distance, transportation costs and other spatial factors influence the location of firms, workers and consumers.

Authors:

Von Thünen, Johann Heinrich: German economist and geographer considered the father of spatial economics.

Christaller, Walter: German geographer who developed the model of "central places".

Analysis of the problem:

Spatial economics has contributed significantly to the development of theoretical models to explain the location of economic activity. However, some critics point out that these models may be too abstract and do not reflect the complexity of real economic decisions.

Development geography: This perspective focuses on the study of economic inequalities and how economic geography can contribute to the development of disadvantaged regions.

Geography of development:

Development geography, which emerged in the 1960s, focuses on the study of economic inequalities and how economic geography can contribute to the development of disadvantaged regions.

Authors:

Myrdal, Gunnar: Swedish economist who introduced the concept of "vicious circles of underdevelopment".

Friedmann, John: American geographer who developed the "uneven spatial development" model.

Analysis of the problem:

Development geography has made important contributions to understanding the causes of poverty and inequality in the world. However, some critics point out that this approach can be too normative and may not adequately consider the perspectives and needs of local communities.

b. Application of the information provided by Economic Geography in the interpretation of the level of economic development at the national, regional and global levels.

The information provided by economic geography is fundamental for interpreting the level of economic development at the national, regional and global levels. By analyzing the distribution of resources, the territorial organization of economic activity and trade flows, it is possible to identify the factors that contribute to the economic development of a country or region.

For example, the abundance of natural resources, the existence of adequate infrastructure, strategic location in the global context and the presence of a skilled labor force are factors that are generally associated with a higher level of economic development.

Economic geography also allows us to identify disparities in economic development between different regions and countries. By analyzing the causes of these inequalities, specific public policies and development strategies can be designed for each case.

At the global level, economic geography is a fundamental tool for understanding global economic trends and their effects on different countries and regions. The analysis of trade flows, foreign investment and the location of multinational companies allows us to identify the opportunities and challenges faced by countries in a globalized world.

Economic geography plays a fundamental role in the interpretation of the level of economic development at the national, regional and global levels. By analyzing the distribution of resources, the territorial organization of economic activity and trade flows, it is possible to identify the factors that contribute to the economic progress of a country or region.

At the national level:

Resource distribution: The abundance of natural resources, such as minerals, water, land and forests, is a key factor for economic development. Economic geography makes it possible to identify the regions with the greatest potential for the exploitation of these resources and to design strategies for their sustainable use.

Infrastructure: Adequate infrastructure, such as roads, ports, airports and energy networks, is essential for transporting goods and services, communication and attracting investment. Economic Geography helps to assess the quality and coverage of infrastructure and to identify areas in need of improvement.

Localization of economic activity: The concentration of industries, urban centers and infrastructure in certain regions can drive economic growth and job creation. Economic geography allows us to analyze the patterns of location of economic activity and to understand the reasons behind these concentrations.

At the regional level:

Economic inequalities: Economic Geography makes it possible to identify disparities in economic development between different regions within a country. By understanding the causes of these inequalities, it is possible to design specific public policies for each region and promote a more balanced development.

Regional potentials: Each region has unique characteristics that can be exploited to boost its economic development. Economic geography helps to identify the potential of each region, such as natural resources, skilled labor or strategic location, and to design strategies for its development.

Regional cooperation: Cooperation between regions can be an effective tool for boosting economic development. Economic Geography can identify opportunities for collaboration between regions and facilitate the coordination of efforts to achieve common goals.

Worldwide:

Globalization: Economic Geography analyzes how globalization affects the spatial organization of economic activity, including the location of multinational firms, foreign investment flows, and patterns of international trade.

Global competitiveness: Countries compete with each other to attract investment, generate exports and participate in global markets. Economic Geography allows us to identify the competitive advantages of each country and develop strategies to strengthen them.

Global inequalities: Disparities in economic development between countries are a major challenge worldwide. Economic Geography helps to understand the causes of these inequalities and to seek solutions to reduce them.

The information provided by Economic Geography is fundamental to interpreting the level of economic development at the national, regional and global levels. By analyzing the geographic and economic factors that influence economic activity, opportunities for economic progress can be identified, inequalities can be addressed, and more sustainable and inclusive development can be promoted.

The interpretation of the level of economic development should not be based solely on economic indicators, but should also consider social, environmental and institutional aspects.

Economic geography is a useful tool for the analysis of economic development, but it is not the only one. It must be complemented by other disciplines such as economics, sociology and political science.

Understanding the level of economic development is essential for designing effective public policies and promoting the well-being of societies.

1.2. Economic geography as a complementary field: A comprehensive analysis.

Economic geography is a discipline that lies at the intersection of geography and economics, studying the spatial distribution of economic activities and how they interact with the geographical environment. In this sense, economic geography becomes an essential complementary field for understanding various aspects of today's world, from international trade and globalization to local development and environmental sustainability.

1. Complementarity with other disciplines:

➢ Economics: Economic geography complements traditional economics by providing a spatial perspective on economic phenomena. This provides an understanding of how geographic location, natural resources, and environmental characteristics influence the production, consumption, and exchange of goods and services.

➢ Geography: Economic geography enriches traditional geography by incorporating economic concepts into the analysis of geographic phenomena. This allows us to understand how economic activities shape the landscape, influence population distribution and generate specific spatial patterns.

➢ Other disciplines: Economic geography complements other disciplines such as sociology, politics and environmental science to provide a comprehensive understanding of the challenges and opportunities facing societies in the globalized world.

2. Contributions of economic geography:

➢ Location analysis: Economic geography helps to understand why certain economic activities are located in certain places and how these locational factors influence their success or failure.

➢ Spatial dynamics of the economy: Economic geography analyzes how economic activities are distributed in space and how this distribution changes over time, under the influence of factors such as globalization, technological development and public policies.

➢ Economic impacts of geographic phenomena: Economic geography evaluates how geographic phenomena, such as natural disasters or climate change, can affect economic activities and the well-being of populations.

> Regional and local development: Economic geography contributes to the design of regional and local development strategies that take into account the geographic characteristics, available resources and economic potential of each territory.

3. Examples of applications:

> Analysis of the location of industries: Economic geography can explain why certain industries are concentrated in specific areas, such as the textile industry in Southeast Asia or the automotive industry in Detroit.

> Study of international trade flows: Economic geography analyzes international trade patterns, considering factors such as distance, transportation costs and trade barriers.

> Assessing the economic impact of climate change: Economic geography can assess how climate change affects agriculture, fisheries and other economic activities in different regions of the world.

> Designing public policies for regional development: Economic geography can inform the formulation of public policies that promote sustainable economic development in rural areas or less favored regions.

Economic geography is emerging as an essential complementary field for understanding today's world and addressing the challenges of the future. Its spatial focus, its ability to integrate concepts from various disciplines, and its wide range of applications make it a valuable tool for public, private, and academic decision-making.

1.3. Classification of economic sectors.

The economy is traditionally divided into four main sectors: primary, secondary, tertiary and quaternary. Each sector comprises a set of economic activities that play a fundamental role in the production, distribution and consumption of goods and services. A detailed explanation of each sector is presented below:

1. Primary sector:

Activities: Covers the extraction of natural resources directly from the environment, such as agriculture, livestock, fishing, forestry and mining.

Characteristics: It is characterized by its direct relationship with nature and its intensive use of natural resources. The products of the primary sector are generally raw materials that serve as inputs for the other economic sectors.

Importance: It is essential for food security, employment generation in rural areas and the provision of raw materials for other industries.

2. Secondary sector:

Activities: Comprises the transformation of raw materials obtained from the primary sector into finished products or consumer goods. Includes activities such as manufacturing, construction and industry.

Characteristics: It is characterized by the use of machinery, technology and production processes to transform raw materials. The secondary sector generates high added value to products and contributes significantly to economic growth.

Importance: It is fundamental for industrial development, the creation of skilled jobs and the economic competitiveness of a country.

3. Tertiary sector:

Activities: Also known as the service sector, it comprises the provision of intangible services to consumers, businesses and other economic sectors. It includes activities such as commerce, transportation, health, education, tourism, finance and professional services.

Characteristics: It is characterized by the predominance of intellectual work over manual labor and the importance of human interaction in the provision of services. The tertiary sector has experienced significant growth in recent decades, becoming one of the main engines of the modern economy.

Importance: It is fundamental to social welfare, quality of life and human development. The tertiary sector generates jobs in various fields and contributes to the satisfaction of the population's needs.

4. Quaternary sector:

Activities: Also known as the knowledge sector, it comprises the production, processing and transmission of information and knowledge. It includes activities such as scientific research, technological development, information and communication technologies (ICT), higher education and specialized consulting.

Characteristics: It is characterized by the high added value of its products and services, its dependence on highly qualified human capital and its significant impact on innovation and economic development.

Importance: It is an emerging sector with great potential for economic growth and competitiveness in the globalized economy. The quaternary sector contributes to the generation of new knowledge, the formation of human capital and the creation of new business opportunities.

5. Interrelationships between sectors:

The four economic sectors do not operate in isolation, but are interconnected and dependent on each other. Raw materials from the primary sector are essential for the secondary sector, which in turn produces goods that are marketed and distributed by the tertiary sector. The quaternary sector generates the knowledge and technology that are used by the other sectors to improve their processes and products.

Understanding economic sectors is fundamental for analyzing the functioning of the economy and its impact on society. Each sector plays a crucial role in the production, distribution and consumption of goods and services, and their interrelationship contributes to a country's economic and social development. As the economy evolves, the importance of the tertiary and quaternary sectors increases, reflecting the growing importance of knowledge, information and innovation in today's world.

CHAPTER 2 Natural Resources: The Basis for Productive Economic Activities and Country Development

2.1. Meaning and identification of the concept of natural resources and the importance of natural resources in economic development.

2.2. Definition of natural resource:

A natural resource is defined as an element or component of the environment that has intrinsic value and can be used to satisfy human needs.

Natural resources are essential elements for the economic development and well-being of nations. Their rational and sustainable use is crucial to ensure a prosperous and environmentally responsible future. This research work analyzes in depth the natural resources, their classification, importance and the role they play in productive economic activities.

These resources are classified into two main groups:

➢ Renewable natural resources: Those that can regenerate naturally, such as water, forests and fauna.

➢ Non-renewable natural resources: Those that exist in limited quantities and cannot be replaced in the short term, such as minerals and fossil fuels.

However, the excessive exploitation of natural resources can have negative consequences for the environment, such as deforestation, pollution and the depletion of water resources. It is therefore essential to promote a rational and sustainable use of these resources, ensuring their conservation for future generations.

2.2.1. Importance of natural resources in economic development:

Natural resources are essential for the economic development of countries. They are the basis for productive activities in sectors such as agriculture, industry and mining. They also provide essential environmental services, such as climate regulation, water purification and oxygen production.

Access to and efficient management of natural resources are key factors for economic growth, job creation and improved well-being of the population. However, overexploitation of these resources can have negative consequences for the environment, such as deforestation, pollution and depletion of water resources.

Some authors have expressed their opinion on the importance of natural resources as follows:

1. Smith, Adam (1776). The Wealth of Nations.

Smith considered natural resources as an essential factor for economic growth. He argued that abundance of natural resources and efficiency in their use were key to the prosperity of nations.

2. Ricardo, David (1817). Principles of political economy and taxation.

Ricardo emphasized the importance of land as a fundamental natural resource. He argued that soil fertility and agricultural productivity were determinants of economic well-being.

3. Malthus, Thomas Robert (1798). Essay on the principle of population.

Malthus warned about the limits of population growth in relation to the availability of natural resources, especially food. His work generated debates on the importance of sustainable resource management and family planning.

4. Jevons, William Stanley (1866). The theory of political economy.

Jevons introduced the concept of "resource economics" and warned of the possibility of the depletion of non-renewable resources such as coal. His work highlighted the need for efficient and sustainable use of resources.

5. Marshall, Alfred (1890). Principles of economics.

Marshall considered natural resources as part of a nation's "capital," along with human capital and physical capital. He argued that investment in the management and development of natural resources was crucial to long-term economic growth.

6. Pigou, Arthur Cecil (1920). The economics of welfare.

Pigou introduced the concept of "externalities" to analyze the environmental and social impacts of natural resource exploitation. He argued that companies should internalize these costs to promote more sustainable economic development.

7. Solow, Robert M. (1957). The Solow growth model.

Solow developed a mathematical model that emphasized the importance of physical capital and human capital for economic growth. However, his model did not explicitly consider the role of natural resources.

8. Brundtland, Gro Harlem (1987). Our Common Future: Report of the World Commission on Environment and Development.

The Brundtland Report, also known as the Bruntland Report, introduced the concept of "sustainable development" as an approach that seeks to balance economic growth with environmental protection and social well-being. The importance of sustainable natural resource management was emphasized in this report.

9. Daly, Herman E. (1992). Steady-state economics: A call for a new paradigm.

Daly proposed steady state economics as an alternative to the model of unlimited economic growth. He advocated a limited use of natural resources and a transition to a more sustainable economy.

10. Stiglitz, Joseph E. (2011). The price of the future: A call to action to save the planet.

Stiglitz discusses market failures and the need for public policies to address environmental challenges, including overexploitation of natural resources and climate change. He emphasizes the importance of economic valuation of ecosystem services and the transition to a green economy.

2.3. Renewable natural resources

2.3.1. Inland waters:

Inland waters, such as rivers, lakes and groundwater, are essential for human life and economic development. They are used for human consumption, agriculture, industry, hydropower generation and navigation.

Importance of inland waters as a renewable natural resource

Inland waters, such as rivers, lakes, lagoons, subway aquifers and wetlands, are a renewable natural resource of vital importance to the planet and to humanity. These waters fulfill essential functions in various fields, including:

Hydrological Cycle:

Inland waters are a fundamental component of the hydrological cycle, which regulates the distribution and movement of water on Earth. Water evaporates from oceans, rivers, lakes, and other surfaces, condenses in the atmosphere, and then precipitates as rain or snow, returning to the land to complete the cycle.

Ecosystem sustainability:

Inland waters are essential for the maintenance of a great diversity of terrestrial and aquatic ecosystems. These waters provide habitat for a wide variety of plants and animals, including fish, amphibians, reptiles, birds and mammals. Wetlands, in particular, are considered one of the most biodiverse ecosystems on the planet.

Fresh water supply:

Inland waters are the main source of freshwater for human consumption, agriculture, industry, and hydropower generation. Freshwater is an indispensable resource for life and human development, and its availability is essential to meet the needs of a constantly growing population.

Climate regulation:

Inland waters play an important role in regulating climate at local and regional levels. These waters absorb and release heat, which helps moderate temperature extremes and regulate precipitation patterns.

Recreation and Tourism:

Inland waters are an important resource for recreation and tourism. Rivers, lakes, and beaches offer opportunities for activities such as swimming, fishing, diving, boating, and hiking. Inland water-related tourism generates significant economic income for many communities around the world.

Resilience to climate change:

Inland waters can be a key resource for climate change adaptation. Freshwater ecosystems can help absorb excess rainwater and reduce the risk of flooding, while wetlands can act as natural barriers against sea level rise.

Inland waters are a renewable natural resource of vital importance for the planet and for humanity. Its conservation and sustainable use are essential to ensure the present and future well-being of generations to come. It is essential to protect freshwater ecosystems, promote sustainable agricultural and urban practices, and manage water responsibly to ensure the availability of this vital resource for all.

Despite being a renewable resource, inland waters are under pressure due to pollution, overexploitation and climate change.

Sustainable inland water management requires a holistic approach that considers human needs, environmental protection and social justice.

The active participation of local communities and civil society in general is essential for the conservation and sustainable use of inland waters.

2.3.2. Fruit resources:

Fruit resources, such as fruits, vegetables and fruit trees, are an important source of food and nutrition for the population. In addition, fruit and vegetable production generates employment and income in rural areas.

2.3.3. Soils:

Soils are a vital natural resource for agriculture, providing support for plants and supplying them with the nutrients necessary for growth. Soil quality is critical to agricultural productivity and food security.

Nutrient cycling in soil: a vital process for life

Soil nutrient cycling, also known as biogeochemical nutrient cycling, is a fundamental process for life on Earth. This cycle involves the transformation and exchange of essential nutrients between living (plants, animals and microorganisms) and non-living (soil, water and air) components of the ecosystem.

Stages of nutrient cycling in soil:

The nutrient cycle in soil can be divided into five main stages:

1. Decomposition:

Organic matter, such as dead leaves, animal remains and plant roots, is decomposed in the soil by microorganisms (bacteria, fungi and other decomposers). These microorganisms release mineral nutrients into the soil.

2. Mineralization:

Complex organic nutrients are converted into simple inorganic forms that plants can absorb. This mineralization process is carried out by soil microorganisms.

3. Absorption:

Plants absorb mineral nutrients from the soil through their roots. These nutrients are essential for plant growth and development.

4. Cycling:

Nutrients circulate within the food chain. Animals obtain the nutrients they need by consuming plants or other animals. Nutrients not used by animals are excreted in the form of urine and feces, returning to the soil.

5. Loss:

Some of the nutrients may be lost from the ecosystem through erosion, leaching (washed out by rainwater) and volatilization (loss of gases to the atmosphere).

Importance of nutrient cycling in soil:

Nutrient cycling in soil is essential for life on Earth for several reasons:

➤ Provides nutrients to plants: Nutrients are essential for plant growth and development. Without nutrient cycling, plants would not be able to survive.

➤ Maintains soil fertility: The decomposition of organic matter and mineralization of nutrients helps maintain soil fertility, making it more productive for plant growth.

➤ Regulates water quality: Nutrient cycling helps regulate water quality by absorbing and filtering contaminants from the soil.

➢ Contributes to biodiversity: Nutrient cycling creates a healthy environment for a wide variety of living organisms, which contributes to ecosystem biodiversity.

Factors affecting nutrient cycling in soil:

Several factors can affect nutrient cycling in soil, including:

➢ Human activities: Intensive agriculture, deforestation and soil pollution can alter nutrient cycling and reduce soil fertility.

➢ Climate change: Climate change can affect water availability and the decomposition of organic matter, which can have a negative impact on nutrient cycling.

➢ Soil erosion: Soil erosion can lead to the loss of essential nutrients, which can affect the productivity of the land.

How to protect nutrient cycling in the soil:

It is important to take measures to protect nutrient cycling in the soil, such as:

➢ Promote sustainable agricultural practices: Adopting sustainable agricultural practices such as organic farming, reduced tillage and crop rotation can help maintain soil fertility and reduce nutrient loss.

➢ Reduce soil pollution: Avoiding excessive use of chemical fertilizers and pesticides, and properly managing agricultural and livestock waste, can help reduce soil pollution and protect nutrient cycling.

➢ Conserve forests: Forests are important for regulating the water cycle and protecting soil from erosion. Conserving forests helps maintain nutrient cycling in the soil.

Nutrient cycling in soil is a vital process for life on Earth. It is essential to protect this cycle by adopting sustainable practices and reducing soil pollution to ensure soil fertility and food availability for present and future generations.

Agrological capacity of soils in panama and their distribution

Panama has a great diversity of soils, with characteristics that vary considerably throughout the national territory. This diversity is due to a combination of factors such as climate, topography, geology and vegetation. The agrological capacity of soils, defined as the aptitude of a soil to support the cultivation of certain agricultural products in a sustainable manner, also presents a wide variation in the country.

Classification of the agrological capacity of soils in Panama:

The Land Classification System for Agrological Capacity (STC) is the method used in Panama to evaluate the agrological capacity of soils. This system classifies soils into eight classes, according to their limitations for agricultural use:

Class I: Soils with the best capacities for agricultural production, with minimal or no limitations. These soils are suitable for a wide range of crops and can support intensive agricultural practices.

Class II: Soils with some limitations for agricultural production, which can be managed with appropriate agricultural practices. These soils are suitable for a variety of crops, but may require conservation measures and special management to maintain their productivity.

Class III: Soils with moderate limitations for agricultural production, which require specific agricultural practices to maintain their productivity. These soils are suitable for less demanding crops or require special conservation measures.

Class IV: Soils with severe limitations for agricultural production, which are only suitable for crops tolerant to soil conditions or for extensive grazing. These soils require careful management practices to avoid degradation.

Class V: Soils unsuitable for agricultural production, due to extreme limitations such as steep slopes, stoniness or permanent waterlogging. These soils can be used for biodiversity conservation or non-agricultural purposes.

Class VI: Soils with extreme limitations for agricultural production, but which can be used for extensive grazing or for the production of firewood or timber. These soils require careful management practices to avoid degradation.

Class VII: Soils unsuitable for agricultural production or extensive grazing, due to extreme limitations such as steep slopes, stoniness or permanent waterlogging. These soils can only be used for biodiversity conservation.

Class VIII: Soils unsuitable for any use, due to extreme limitations such as steep slopes, stoniness or permanent waterlogging. These soils can only be used for biodiversity conservation.

Distribution of the agrological capacity of soils in Panama:

The distribution of the agrological capacity of soils in Panama is uneven. The areas with the greatest agricultural potential are found in the Pacific plains and in some areas of the interior of the country, such as the Changuinola Valley. These areas are dominated by Class I, II and III soils, which are suitable for a variety of crops.

The areas with the least agricultural potential are located in the mountainous areas of the country, where soils of classes IV, V, VI and VII predominate. These soils are less suitable for agricultural production and require careful management practices to avoid degradation.

Importance of the Agrological Capacity of Soils:

Information on the agrological capacity of soils is fundamental for land use planning and for the development of sustainable agriculture in

Panama. This information makes it possible to identify the areas with the greatest potential for agricultural production and to make informed decisions on crop selection, agricultural practices and necessary conservation measures.

Panama has a great diversity of soils with different agrological capacities. The distribution of these capacities in the national territory is uneven, with areas of greater and lesser agricultural potential. Information on the agrological capacity of soils is essential for land use planning and for the development of sustainable agriculture in the country.

The agrological capacity of soils is not a static factor, but can change over time due to human action or natural factors such as erosion or climate change.

It is important to use sustainable agricultural practices that conserve soil fertility and productivity.

Land use planning must consider the agrological capacity of soils to avoid environmental degradation and loss of productivity.

2.4. Non-renewable natural resources

Non-renewable natural resources are those that, once depleted, cannot be replaced on a human scale within a reasonable geological time. Unlike renewable resources, such as water or forests, which can regenerate naturally, non-renewable resources are finite and their exploitation entails significant environmental and economic challenges.

Importance of non-renewable natural resources:

Non-renewable resources have played a fundamental role in the development of modern society. They are essential for industry, transportation, power generation and the manufacture of a wide range of products.

Industry: Minerals are used as raw materials in the production of steel, aluminum, cement and other construction materials.

Transportation: Fossil fuels are the main source of energy for land, air and maritime transportation.

Power generation: Coal, oil and natural gas are used to generate electricity in thermal power plants.

Product manufacturing: Non-renewable resources are used in the manufacture of a wide range of products, from cell phones to computers to automobiles.

2.5. The mining resource:

Mineral resources, such as iron, copper, gold and oil, are essential to modern industry. They are used in the manufacture of machinery, equipment, vehicles and other products.

2.5.1. Most used minerals and their commercialization:

The most widely used minerals in the world are:

➢ Iron: Used in construction, automobile manufacturing and other products.

➢ Oil: Used as fuel for transportation, power generation and plastics production.

➢ Natural gas: Used as a fuel for power generation and heating.

➢ Copper: Used in the manufacture of electrical cables, pipes and other products.

➤ Gold: Used in jewelry, electronics and as a store of value.

Minerals are traded through international markets, where prices are set according to supply and demand. Countries with large mineral reserves can obtain significant income from their export.

2.5.2. Mineral resource depletion and consequences:

The excessive exploitation of mineral resources can lead to their depletion, which would have serious consequences for the economy and the well-being of the population. Mineral depletion can lead to:

➤ Rising prices: As mineral resources become scarcer, their prices tend to rise.

➤ Energy insecurity: Dependence on fossil fuels, such as oil and natural gas, can lead to instability in energy supply.

➤ Environmental degradation: Mining can have a negative impact on the environment, such as deforestation, water pollution and soil erosion.

Challenges in the exploitation of non-renewable natural resources:

The excessive exploitation of non-renewable resources brings with it significant challenges:

➤ Depletion: The extraction of these resources at an accelerated rate may lead to their depletion in the relatively near future.

➤ Environmental impacts: Mining and fossil fuel extraction can generate considerable environmental damage, such as water and soil contamination, deforestation and greenhouse gas emissions.

➤ Energy dependence: Dependence on fossil fuels for energy

generation makes us vulnerable to price fluctuations and geopolitical risks.

Sustainability and alternatives:

It is essential to adopt a sustainable approach to the management of non-renewable natural resources. This implies:

➢ Reduce consumption: Implement energy efficiency measures and promote the use of renewable energies.

➢ Recycle and reuse: Encourage the recycling of materials from non-renewable resources to reduce the demand for new resources.

➢ Develop alternatives: Invest in research and development of technologies that allow the use of renewable energy sources and alternative materials.

2.6. Mining activity in Panama

History of Mining in Panama: A Journey Through Time

Panama, located in the heart of the Central American isthmus, has a rich mining history dating back to pre-Columbian times. From the extraction of gold by indigenous people to the exploitation of copper and other minerals today, mining has played a significant role in the country's economic and social development.

Origins of mining in Panama:

The first evidence of mining activity in Panama dates back thousands of years. The indigenous people, mainly the Ngobe-Buglé and the Guna, extracted gold from rivers and streams using rudimentary techniques such as sand washing and the use of pans. Gold was used for the manufacture of ornaments, ceremonial tools and as a medium of exchange.

Arrival of the Spaniards and large-scale exploitation:

The arrival of the Spanish conquistadors in the 16th century marked a turning point in Panama's mining history. The Spaniards, attracted by tales of gold riches, established settlements and exploited the mines intensively, using indigenous and slave labor. The gold mined in Panama was sent to Spain, contributing significantly to the economy of the Spanish Empire.

Decline of colonial mining and boom in the 19th century:

Colonial mining had a devastating impact on indigenous populations and the environment. The overexploitation of resources, the diseases brought by the Spaniards and the violence exerted on the indigenous people caused a drastic decrease in the indigenous population and a decline in mining activity at the end of the 18th century.

However, mining experienced a resurgence in the 19th century with the advent of the California Gold Rush and the construction of the Panama Transisthmian Railroad. The railroad route facilitated the transport of people and goods between the Atlantic and Pacific oceans, which boosted the search for gold in the areas surrounding the railroad.

Modern mining and current challenges:

Mining in Panama is currently concentrated mainly in the exploitation of copper, molybdenum and other minerals. The mining industry has generated significant revenues for the country and has contributed to infrastructure development and job creation. However, it has also generated controversy due to its environmental and social impacts.

Environmental impacts of mining:

Mining activity can generate various environmental impacts, such as deforestation, water and soil contamination, toxic waste generation and landscape alteration. These impacts can affect biodiversity, the health of local communities, and the quality of drinking water.

Social impacts of mining:

Mining can also generate negative social impacts, such as the displacement of communities, human rights violations, social conflict and the precariousness of labor. It is important that mining companies implement measures to mitigate these impacts and ensure respect for human rights and the environment.

Future of mining in Panama:

The future of mining in Panama will depend on the country's ability to strike a balance between economic development, environmental protection and social welfare. It is essential that the mining industry operates in a responsible, sustainable and transparent manner, and that the participation of local communities in mining-related decision-making is guaranteed.

2.6.1. Main mining areas of the country:

Panama's main mining areas are located in the provinces of Colon, Veraguas, Chiriqui and Bocas del Toro. The most exploited minerals in the country are copper, gold, silver and bauxite.

2.6.2. National regulation of mining activities in Panama:

Mining activity in Panama is regulated by the Mining Code and other laws and regulations.

National Regulation of Mining Activity in Panama: A Complex Legal Framework Mining activity in Panama is regulated by a complex legal framework involving various laws, decrees, regulations and technical standards. The main objective of this regulation is to guarantee responsible and sustainable mining development that respects the environment and local communities.

Main laws regulating mining in Panama:

1. Mineral Resources Code (Decree Law 23 of 1963): This law

establishes the general principles governing mining activity in Panama, including the rights and obligations of mining companies, the granting of mining concessions and environmental protection.

2. Law 11 of 1997: This law created the National Environmental Authority (ANAM), which is responsible for environmental protection and natural resource management. ANAM plays a key role in regulating mining activity, establishing environmental standards and granting environmental permits for mining projects.

3. Law 69 of 2009: This law creates the Ministry of the Environment (MiAmbiente), which assumes the functions of ANAM in environmental protection. MiAmbiente continues to be a key entity in the regulation of mining activities.

4. Law 12 of 2013: This law creates the National Authority for Transparency and Access to Information (ANTAI), which is responsible for guaranteeing access to public information, including information related to mining activity. ANTAI plays an important role in promoting transparency and accountability in the mining sector.

Other relevant laws and regulations:

1. Executive Decree 124 of 2014: This decree regulates the procedure for obtaining environmental permits for mining projects.

2. Regulation for the Prevention and Control of Air Pollution by Emissions from Fixed Sources: This regulation establishes the maximum permissible limits of air pollutant emissions for mining

projects.

3. Environmental Technical Standard for the Management of Solid Hazardous Waste: This standard establishes the requirements for the handling, transportation and final disposal of solid hazardous waste generated by mining activities.

Institutions responsible for mining regulation:

1. Ministry of Commerce and Industries (MICI): This is the entity responsible for granting mining concessions and supervising the companies' compliance with mining regulations.

2. Ministry of Environment (MiAmbiente): This is the entity responsible for ensuring environmental protection and granting environmental permits for mining projects.

3. National Authority for Transparency and Access to Information (ANTAI): It is the entity responsible for guaranteeing access to public information related to mining activities.

Challenges of mining regulation in Panama:

The regulation of mining activity in Panama faces several challenges, such as:

1. Institutional weaknesses: Fragmentation of responsibilities among different government entities can hinder effective enforcement.

2. Lack of transparency: Lack of access to public information on mining projects can generate distrust in local communities and hinder citizen

participation in decision-making processes.

3. Corruption: Corruption can affect transparency and accountability in
    the mining sector, which can lead to improper exploitation of natural
    resources and violation of human rights.

The regulation of mining activity in Panama is a complex issue that requires a comprehensive approach that considers economic, environmental, social and institutional aspects. It is essential to strengthen the institutions responsible for mining regulation, promote transparency and citizen participation, and combat corruption to ensure sustainable and responsible mining development in the country.

2.5.3. Effects of mining activities on the environment: an analysis based on environmental impact assessments (EIA).

Mining activity, while an important source of economic resources for many countries, can have a significant impact on the environment. These impacts can be both short and long term, and can affect various components of the ecosystem, including soil, water, air, flora and fauna.

Effects on soil:

1. Soil erosion and degradation: Removal of topsoil during open-pit
    mining can lead to erosion, loss of fertility and desertification.

2. Soil contamination: Mined minerals, chemicals used in mineral
    processing and waste generated by mining activities can
    contaminate the soil with heavy metals, acids and other harmful
    substances.

3. Alteration of soil structure: Soil compaction caused by the movement
    of heavy machinery can alter its structure, hindering water infiltration
    and air circulation.

Effects on water:

1. Water contamination: Mining waste, chemical spills and acid mine drainage can contaminate surface and groundwater with heavy metals, acids and other harmful substances.

2. Decreased water availability: Mining activity can consume large amounts of water, which can reduce the availability of water for other uses, such as agriculture and human consumption.

3. Alteration of water flow: The construction of mining infrastructure, such as dams and canals, can alter the natural flow of water, affecting aquatic ecosystems and the local communities that depend on it.

Effects in the air:

1. Air pollution: Mining activity generates emissions of dust, gases and other pollutants that can affect air quality, especially in communities near the mines.

2. Acid rain: The emission of gases such as sulfur dioxide and nitrogen oxide during mining activity can contribute to the formation of acid rain, which damages forests, crops and aquatic ecosystems.

3. Climate change: The burning of fossil fuels for power generation used in mining contributes to greenhouse gas emissions, which accelerates climate change.

Effects on flora and fauna:

1. Habitat loss: Deforestation and destruction of natural habitats for mine construction and mine waste disposal can affect local flora and fauna, leading to species loss and ecosystem fragmentation.

2. Contamination of the food chain: Animals that consume water or plants contaminated by mining activity can accumulate toxins in their bodies, affecting their health and reproducing.

3. Alteration of animal behavior: Noise, vibration and artificial light generated by mining activities can alter the behavior of local fauna, affecting their ability to feed, reproduce and migrate.

Mitigation of environmental impacts of mining:

It is important to implement measures to mitigate the environmental impacts of mining activities, including:

1. Environmental impact studies: Conduct environmental impact studies prior to the opening of a mine to identify and assess potential environmental impacts and develop appropriate mitigation plans.

2. Clean technologies: Implement clean and efficient technologies for the extraction, processing and transportation of minerals, in order to reduce the generation of waste and polluting emissions.

3. Environmental restoration: Implement environmental restoration plans to recover areas affected by mining activities, reforesting land, rehabilitating soils and reintroducing native species.

4. Community participation: Involve local communities in the planning and execution of mining projects, ensuring their participation in

decision making and benefit sharing.

Mining activity can have a significant impact on the environment, affecting soil, water, air, flora and fauna. It is essential to implement measures to mitigate these impacts, using clean technologies, restoring affected areas and ensuring the participation of local communities. Responsible mining must seek a balance between the exploitation of mineral resources and environmental protection to ensure sustainable development.

The severity of the environmental impacts of mining activity varies according to the type of mining, the geology of the terrain, the practices used and the mitigation measures implemented. Regulatory frameworks and policies are essential.

CHAPTER 3 Energy and its Importance in the National Economy

The history of energy use is closely linked to the development of mankind. In the early days, people relied on renewable energy sources such as wood, moving water and solar energy. The Industrial Revolution marked a turning point with the development of technologies to harness non-renewable energy sources such as coal and oil. These energy sources enabled unprecedented economic growth, but also generated significant environmental impacts. Today, the search for sustainable and efficient energy sources is a crucial challenge for sustainable development.

3.1. Historical evolution of energy use

From prehistoric times to the present day. We will see how energy has been fundamental to the development of mankind, driving production, transportation and quality of life.

3.1.1. Before the Industrial Revolution

Energy came mainly from renewable sources such as firewood, water and wind.

The energy was mainly used for domestic tasks, such as cooking and heating homes.

Some ancient civilizations developed technologies to harness solar and wind energy, such as concave mirrors to concentrate the sun's heat or windmills to grind grain.

Before the Industrial Revolution, energy came mainly from renewable sources such as wood, water and wind. These energy sources were limited and difficult to control, which restricted economic and technological development. Firewood was the main source of energy for cooking and heating homes, while water power was used to grind grain and pump water. Some ancient civilizations developed technologies to harness solar and wind energy, such as concave mirrors to concentrate the sun's heat or windmills to grind grain. However, these technologies were rudimentary and did not provide enough energy to fuel economic and social growth.

3.1.2. The industrial revolution and changes in production levels

The Industrial Revolution marked a turning point in the use of energy.

New technologies were developed that made it possible to take advantage of non-renewable energy sources such as coal and oil.

These energy sources provided greater power and control, which drove a significant increase in industrial production and transportation.

The Industrial Revolution marked a turning point in the use of energy. New technologies were developed that made it possible to harness non-renewable energy sources such as coal and oil. These energy sources provided greater power and control, which drove a significant increase in industrial production and transportation. The availability of energy on a large scale enabled the development of more efficient machines, the growth of cities and the expansion of international trade. The steam engine, invented by James Watt in 1769, was a key milestone in the Industrial Revolution. This engine enabled the conversion of thermal energy from coal into mechanical energy, which could be used to power factories, trains and ships. The development of the internal combustion engine in the late 19th and early 20th centuries marked another important advance in the use of energy. This engine enabled the development of automobiles, airplanes and other vehicles that revolutionized transportation.

Impacts of the Industrial Revolution on energy use

Increased energy consumption: The Industrial Revolution led to an exponential increase in energy consumption, driven by the growing demand for energy for industrial production, transportation and home heating.

Development of new technologies: The Industrial Revolution spurred the development of new technologies for the extraction, conversion and distribution of energy, such as the steam engine, the internal combustion engine and the electric grid.

Changes in economic structure: The availability of large-scale energy contributed to the transition from an agrarian to an industrial economy, with a greater emphasis on manufacturing and trade.

Environmental impacts: The use of non-renewable fossil fuels during the Industrial Revolution generated significant environmental impacts, such as air pollution, acid rain and climate change.

The Industrial Revolution had a profound impact on global energy use. Energy consumption increased exponentially due to the growing demand for energy for industrial production, transportation and home heating. The development of new technologies, such as the steam engine, the internal combustion engine and the electric grid, made it possible to harness non-renewable energy sources such as coal and oil more efficiently. These changes spurred the transition from an agrarian to an industrial economy, with a greater emphasis on manufacturing and commerce. However, the use of non-renewable fossil fuels during the Industrial Revolution also generated significant environmental impacts, such as air pollution, acid rain and climate change. These environmental impacts remain a major challenge today and require urgent action to mitigate and reduce them.

3.2. The different types of energy and their current use

Energy is classified into two main categories: renewable and non-renewable.

Renewable energy sources are naturally replenished, such as solar, wind, hydro and geothermal energy.

Non-renewable energy sources are finite, such as coal, oil and natural gas.

Today, various types of energy are used to meet society's needs. Renewable energy sources are gaining increasing importance due to their sustainability and lower environmental impact. However, non-renewable energy sources remain predominant in the global energy matrix, raising concerns about resource depletion and greenhouse gas emissions.

3.2.1. Of water in motion

Hydroelectric power harnesses the power of moving water to generate electricity.

It is a renewable, clean and reliable source of energy.

It is mainly used to generate electricity on a large scale.

Hydropower is one of the most important renewable energy sources in the world.

It is obtained by harnessing the power of moving water to turn turbines that generate electricity. It is a clean and reliable source of energy that does not produce polluting emissions. However, the construction of hydroelectric dams can have negative environmental impacts, such as flooding natural habitats and altering the flow of rivers.

3.2.2. From hard coal or stone coal

Coal is a fossil fuel used primarily to generate electricity and heat.

It is a non-renewable and polluting source of energy.

Their use contributes to greenhouse gas emissions and climate change.

Coal is a fossil fuel that has been used for centuries to generate electricity and heat.

It is a non-renewable and polluting source of energy, since its combustion releases greenhouse gases and other atmospheric pollutants. Coal use has been associated with respiratory health problems and environmental degradation. Despite its negative impacts, coal remains an important source of energy in many countries, especially those with large reserves of this resource.

3.2.3. Oil

Petroleum is a liquid fossil fuel used primarily for transportation and energy production.

It is a non-renewable energy source

Major oil areas of the world: current situation

The oil industry is one of the most important in the world, with a major influence on the global economy and geopolitics. Oil is a non-renewable natural resource used primarily as a fuel for transportation, power generation and the production of petrochemicals.

Distribution of oil reserves:

Oil reserves are unevenly distributed around the world. The main countries with proven oil reserves are:

Venezuela: With 200 billion barrels, Venezuela has the largest proven oil reserves in the world. However, the country's oil production has declined significantly in recent years due to the economic and political crisis.

Saudi Arabia: With 174 billion barrels, Saudi Arabia is the country with the second largest proven oil reserves. It is one of the world's leading oil producers and exporters.

Canada: With 169 billion barrels, Canada has the third largest proven oil reserves in the world. Canada's oil reserves are found mainly in the Alberta oil sands.

Russia: With 158 billion barrels, Russia is the country with the fourth largest proven oil reserves. It is a major producer and exporter of oil and natural gas.

United States: With 135 billion barrels, the United States is the country with the fifth largest proven oil reserves. It has become the world's largest oil producer thanks to the development of the fracking technique.

Current situation of the oil industry:

The oil industry faces several challenges today:

- Energy transition: The world is in transition to a low-carbon economy, which means that demand for oil could decrease in the future. This could have a negative impact on oil-producing countries.

- Climate change: Climate change is generating greater awareness of the need to reduce greenhouse gas emissions, which could lead to lower demand for fossil fuels such as oil.

- Geopolitical instability: Geopolitical instability in some regions of the world may affect oil production and supply, which may generate volatility in crude oil prices.

➤ Development of new technologies: The development of new technologies, such as electric vehicles and renewable energies, could reduce dependence on oil in the future.

Despite these challenges, the oil industry remains an important industry with a significant role in the global economy. The future of the oil industry will depend on its ability to adapt to market changes and new technologies.

The distribution of oil reserves may change over time as new fields are discovered or existing fields are depleted.

The current situation of the oil industry is complex and subject to multiple factors that may affect its evolution in the future.

The energy transition and the fight against climate change are major challenges that the oil industry must address to ensure its long-term sustainability.

3.2.4. Wind

Wind energy harnesses the power of the wind to generate electricity. It is a renewable, clean and abundant source of energy.

It is mainly used in large-scale wind farms.

Wind energy is a renewable energy source that is gaining more and more importance nowadays. It is obtained by harnessing the power of the wind to turn turbines that generate electricity. It is a clean and abundant source of energy that does not produce polluting emissions. Wind power technology has undergone major development in recent years, which has made it possible to reduce costs and increase the efficiency of wind turbines. Large-scale wind farms are being installed in many parts of the world, contributing to the diversification of the energy matrix and the reduction of greenhouse gas emissions.

Wind energy in Panama: Wind farm areas, social and economic benefits

Panama has experienced significant growth in the development of wind energy in recent years, positioning itself as one of the leaders in the Central American region. Wind energy has become a crucial renewable energy source for the country, contributing to the diversification of the energy matrix, reducing dependence on fossil fuels and mitigating climate change.

Wind farm areas:

The main sites for the installation of wind farms in Panama are located in:

Azuero Peninsula: This region has favorable climatic conditions, with strong and constant winds, which makes it an ideal place for wind energy generation. Currently, several wind farms are operating in the Azuero Peninsula, such as the Penonomé Wind Farm and the Cañafístula Wind Farm.

Isthmus of Panama: The Isthmus of Panama also offers great potential for wind energy, especially in the mountainous areas. Projects are being developed for the construction of wind farms in this region, such as the Changuinola Wind Farm and the Penonomé II Wind Farm.

Social and economic benefits:

Wind energy in Panama generates important social and economic benefits, including:

- ➤ Diversification of the energy matrix: Wind energy reduces the country's dependence on imported fossil fuels, contributing to energy security and national independence.

➢ Reduced greenhouse gas emissions: Wind energy produces no polluting emissions during operation, which helps mitigate climate change and improve air quality.

➢ Job creation: The construction, operation and maintenance of wind farms generate direct and indirect jobs in local communities.

➢ Local economic development: Investment in wind energy boosts the economic development of the regions where wind farms are located, attracting investment and generating business opportunities.

➢ Reduced energy costs: In the long term, wind energy can contribute to reduced energy costs for consumers, especially when compared to volatile fossil fuel prices.

Challenges and opportunities:

Despite the benefits of wind energy, there are some challenges that must be addressed to continue its sustainable development in Panama:

➢ Wind variability: Wind power generation depends on the strength and direction of the wind, which can lead to intermittency in electricity production.

➢ Environmental impacts: The construction of wind farms may have some environmental impacts, such as loss of natural habitat and noise generated by wind turbines.

➢ Investment needs: Initial investment in wind energy projects is significant, which requires a favorable regulatory framework and adequate financing.

However, these barriers can be overcome by implementing appropriate strategies, such as developing energy storage technologies, conducting rigorous environmental impact studies, and creating attractive financing mechanisms for wind energy investment.

Wind energy in Panama represents a significant opportunity for the sustainable development of the country, contributing to the diversification of the energy matrix, the reduction of greenhouse gas emissions and the generation of jobs and economic opportunities. By addressing existing challenges and taking advantage of opportunities, Panama can continue to position itself as a leader in wind energy production and utilization in the region.

3.2.5. Nuclear

Nuclear power harnesses the energy released in nuclear reactions to generate electricity.

It is a controversial energy source due to the risks of accidents and nuclear waste management.

It is mainly used to generate electricity on a large scale.

Nuclear energy is a controversial source of energy that generates global debate. It is obtained through the fission or fusion of atomic nuclei, releasing a large amount of energy that is used to generate electricity. Although nuclear energy does not produce polluting emissions during operation, there are risks associated with nuclear accidents and the management of high-level nuclear waste. Nuclear technology has seen significant advances in safety, but public perception of nuclear power remains complex and requires an open and transparent dialogue.

This energy has been used for a variety of purposes, both peaceful and military. Its use has generated important benefits, such as large-scale electricity generation and the development of medical applications, but it has also had negative consequences, such as nuclear accidents and the proliferation of nuclear weapons.

Cases of nuclear energy use:

Electricity generation: The main application of nuclear power is large-scale electricity generation. Nuclear power plants produce electricity by fission or fusion of atomic nuclei, releasing a large amount of energy that is used to turn turbines and generate electricity.

Medical applications: Nuclear energy is also used in a variety of medical applications, such as radiotherapy for cancer treatment, production of radioisotopes for medical diagnosis and treatment, and sterilization of medical equipment.

Vehicle propulsion: Nuclear power has been used in the past for the propulsion of submarines and some prototype ships and aircraft. However, this application has not been widely developed due to the risks and technical challenges involved.

➢ Nuclear weapons: Nuclear energy has also been used for the development of nuclear weapons, which have enormous destructive power and pose a serious threat to global security and world peace.

Consequences of the use of nuclear energy:

➢ Nuclear accidents: Nuclear accidents are serious events that can have devastating consequences for human health and the environment. Some of the best known nuclear accidents are Chernobyl in 1986 and Fukushima in 2011.

➢ Proliferation of nuclear weapons: The proliferation of nuclear weapons is the expansion of the capability to produce and use nuclear weapons to more countries. This increases the risk of nuclear war, which could have catastrophic consequences for humanity.

➢ Nuclear waste: The production of nuclear energy generates radioactive nuclear waste that is difficult to store and manage safely. These wastes can contaminate the environment and represent

a risk to human health for thousands of years.

Countries currently using nuclear energy:

Currently, about 30 countries in the world use nuclear power for electricity generation. Some of the main countries with nuclear energy are:

- China: China is the country with the largest installed nuclear power capacity in the world.

- United States: The United States is the country with the second largest installed nuclear power capacity in the world.

- France: France is the country with the third largest installed nuclear power capacity in the world.

- Russia: Russia has the fourth largest installed nuclear power capacity in the world.

- South Korea: South Korea has the fifth largest installed nuclear power capacity in the world.

Global health effects:

The effects of nuclear power on global health depend largely on how the technology is used. The generation of electricity through nuclear power produces no polluting emissions during normal operation, which means that it does not contribute to air pollution or climate change. However, nuclear accidents can have serious consequences for human health, as exposure to radiation can cause cancer, heart disease and other health problems.

In addition, the management of radioactive nuclear waste also poses a risk to human health in the long term, as this waste can contaminate water, soil and air and expose people to radiation.

Nuclear power is a powerful energy source that can have both benefits and risks. It is important to use this technology responsibly and safely to minimize risks to human health and the environment. The international community must work together to prevent the proliferation of nuclear weapons and ensure the safe management of nuclear waste.

3.2.6. Atomic

Atomic energy is a type of nuclear energy that is mainly used for military purposes.

It relies on nuclear fission to release a large amount of energy in the form of an explosion.

Their use has serious ethical and humanitarian implications due to their destructive potential.

Atomic energy is a type of nuclear energy that is mainly used for military purposes. It relies on nuclear fission to release a large amount of energy in the form of an explosion. Its use has serious ethical and humanitarian implications due to its destructive potential. Nuclear weapons represent a serious threat to global security and world peace. Nuclear proliferation and the risk of nuclear war are issues of great concern to the international community. It is essential to promote nuclear disarmament and non-proliferation to avoid the catastrophic consequences of a nuclear conflict.

Countries currently using atomic energy: Specific uses and implications for human health:

Atomic energy, also known as nuclear energy, is a controversial energy source that has been used for a variety of purposes since its discovery in the mid-20th century. Its use has generated important benefits, such as large-scale electricity generation and the development of medical applications, but it has also had negative consequences, such as nuclear accidents and the proliferation of nuclear weapons.

Countries currently using atomic energy:

Currently, about 30 countries in the world use nuclear power for electricity generation. Some of the main countries with nuclear energy are:

- China: China is the country with the largest installed nuclear power capacity in the world, with 54 reactors in operation and 18 under construction. Nuclear power accounts for about 5% of China's total electricity generation.
- United States: The United States is the country with the second largest installed nuclear power capacity in the world, with 93 operating reactors. Nuclear power accounts for about 20% of total U.S. electricity generation.
- France: France has the third largest installed nuclear power capacity in the world, with 56 reactors in operation. Nuclear power accounts for about 70% of total electricity generation in France.
- Russia: Russia has the fourth largest installed nuclear power capacity in the world, with 37 operating reactors. Nuclear power accounts for about 17% of Russia's total electricity generation.
- South Korea: South Korea has the fifth largest installed nuclear power capacity in the world, with 24 operating reactors. Nuclear power accounts for about 25% of total electricity generation in South

Korea.

Specific uses of atomic energy:

The main application of nuclear power is large-scale electricity generation. Nuclear power plants produce electricity by fission or fusion of atomic nuclei, releasing a large amount of energy that is used to turn turbines and generate electricity.

In addition to electricity generation, nuclear energy is also used in various medical applications, such as radiotherapy for cancer treatment, production of radioisotopes for medical diagnosis and treatment, and sterilization of medical equipment.

To a lesser extent, nuclear power has been used in the past for vehicle propulsion, such as submarines and some prototype ships and aircraft. However, this application has not been widely developed due to the risks and technical challenges involved.

Consequences for human health:

The effects of nuclear power on global health depend largely on how the technology is used. Nuclear power generation does not produce polluting emissions during normal operation, which means that it does not contribute to air pollution or climate change. However, there are risks associated with nuclear power that can affect human health:

Nuclear accidents: Nuclear accidents are serious events that can have devastating consequences for human health and the environment. Some of the best known nuclear accidents are Chernobyl in 1986 and Fukushima in 2011. These accidents exposed thousands of

people to radiation, causing cancer, heart disease and other long-term health problems.

Radiation exposure: Exposure to ionizing radiation, either from a nuclear accident or from contact with radioactive materials, can cause a variety of health problems, including cancer, heart disease, damage to the reproductive system and developmental problems in children.

Nuclear waste management: Nuclear energy production generates radioactive nuclear waste that is difficult to store and manage safely. This waste can contaminate water, soil and air and expose people to radiation.

Nuclear power is a powerful energy source that can have both benefits and risks. It is important to use this technology responsibly and safely to minimize risks to human health and the environment. The international community must work together to prevent the proliferation of nuclear weapons and ensure the safe management of nuclear waste.

3.2.7. Biomass.

Biomass is an organic material used to generate energy.

Includes wood, agricultural residues and other organic materials.
It is mainly used to generate electricity and heat.

Biomass is a renewable energy source obtained from organic materials such as wood, agricultural residues, forestry waste and energy crops. Biomass can be used to generate electricity, heat and biofuels. It is a relatively clean source of energy, but its use can have environmental impacts if not managed sustainably. Large-scale biomass production can lead to deforestation, loss of biodiversity and greenhouse gas emissions. It is important to promote sustainable practices for the production and use of biomass to ensure its effective contribution to the energy transition.

Biomass has become an increasingly important source of renewable energy in today's world. It is organic material, such as wood, agricultural residues and energy crops, which can be used to generate electricity, heat and biofuels. The use of biomass offers a number of advantages over traditional fossil fuels, including reduced greenhouse gas emissions, improved energy security and job creation.

Countries using biomass:

The use of biomass is widespread throughout the world, with countries using it to varying degrees. Some of the major biomass consuming countries are:

- European Union: The European Union is the largest consumer of biomass in the world, with use accounting for about 10% of total energy consumption.

- United States: The United States is the second largest consumer of biomass in the world, with use accounting for about 5% of total energy consumption.

- Brazil: Brazil is the third largest consumer of biomass in the world, with use accounting for about 40% of total energy consumption.

- China: China is the fourth largest consumer of biomass in the world, with use accounting for about 10% of total energy consumption.

- India: India is the fifth largest consumer of biomass in the world, with

a

use, which accounts for about 30% of total energy consumption.

Biomass applications:

Biomass has a wide range of applications, including:

- Electricity generation: Biomass can be used to fuel power plants that generate electricity from the steam produced by the combustion of organic material.

- Heating: Biomass can be used to heat homes and buildings through direct combustion or in biomass boilers.

- Biofuel production: Biomass can be used to produce biofuels such as biodiesel and ethanol, which can be used for transportation and other applications.

- Chemicals: Biomass can be used to produce a variety of chemical products.

of chemicals, such as plastics, resins and fertilizers.

Health effects:

Biomass can have some negative health effects if not used in a sustainable manner. Biomass combustion can emit air pollutants, such as fine particulate matter and volatile organic material, which can affect air quality and respiratory health.

However, biomass can also have some health benefits. For example, using biomass for heating can help reduce dependence on fossil fuels, which can lead to better air quality and a lower incidence of respiratory diseases.

Advantages of using biomass:

The use of biomass offers a number of advantages over traditional fossil fuels, including:

- ➢ Reducing greenhouse gas emissions: Biomass is a renewable energy source that releases no net greenhouse gases into the atmosphere, which can help mitigate climate change.

- ➢ Improved energy security: Biomass is a domestic energy source that can reduce dependence on fossil fuel imports, which can improve a country's energy security.

> Job creation: The biomass industry creates jobs in a variety of sectors, including agriculture, forestry, manufacturing and construction.

> Rural development: Biomass production can provide income and development opportunities for rural communities.

> Sustainability: Biomass can be a sustainable energy source if managed responsibly, ensuring that the harvest rate does not exceed the natural renewal rate.

Biomass is a versatile and sustainable renewable energy source that offers a number of benefits over traditional fossil fuels. Its use can help reduce greenhouse gas emissions, improve energy security, create jobs and promote rural development. As the world seeks cleaner and more sustainable energy sources, biomass will play an increasingly important role in the energy future.

3.2.8. Geothermal.

Geothermal energy harnesses the Earth's internal heat to generate electricity.

It is a renewable, clean and reliable source of energy.

It is mainly used to generate electricity on a large scale.

Geothermal energy is a renewable energy source that harnesses the Earth's internal heat to generate electricity. It is obtained by extracting steam or hot water from deep wells, which is then used to turn turbines that generate electricity. Geothermal energy is a clean and reliable source of energy that does not produce polluting emissions. However, its development can have environmental impacts, such as greenhouse gas emissions, landscape alteration and water consumption. It is important to conduct environmental impact studies and develop geothermal projects in a responsible manner to minimize these impacts.

Potential negative impacts of geothermal energy

While geothermal energy is considered a renewable and clean energy source, its use can also have some negative environmental and social impacts. It is important to consider these aspects for a sustainable development of geothermal energy.

Environmental impacts:

- Gas emissions: The extraction of steam and hot water from underground can release gases such as carbon dioxide, methane, and hydrogen sulfide into the atmosphere. While emissions from geothermal energy are generally low compared to fossil fuels, they can still contribute to air pollution and the greenhouse effect, especially in geothermal plants with high levels of non-condensable gases.

- Water disturbance: Extracted geothermal water may contain minerals and contaminants that, if not properly treated before reinjection, can pollute surface and groundwater. It is crucial to implement efficient water treatment systems to minimize this impact.

- Seismicity: The injection of water into the subsoil to generate steam can induce small earthquakes in some areas. Although these earthquakes are usually mild, in regions with high seismic activity it is necessary to carry out careful studies and apply measures to prevent or mitigate induced earthquakes.

- Noise and vibrations: The operation of some geothermal plants can generate noise and vibrations that can affect nearby communities. It is important to implement noise and vibration control measures to minimize these nuisances.

- Visual impact: The construction of geothermal plants and cooling towers can have a visual impact on the landscape, especially in sensitive or aesthetically valuable areas. It is important to consider the design and location of the facilities to minimize this impact.

Social impacts:

- Access to water: Geothermal water extraction can affect the availability of water for other uses, such as agriculture or human consumption, especially in areas with water scarcity. It is crucial to conduct water impact studies and establish water management plans to ensure equitable and sustainable use of the resource.

- Land rights: The development of geothermal projects can generate conflicts over land use, especially with indigenous or local communities that depend on the land for their livelihoods. It is essential to establish transparent dialogue and participation processes that ensure respect for land rights and fair distribution of benefits.

- Occupational Hazards: Workers in the geothermal industry may be exposed to occupational hazards such as accidents, exposure to noxious gases, and extreme working conditions. Adequate occupational health and safety measures must be implemented to protect workers.

Mitigation of negative impacts:

To minimize the negative impacts of geothermal energy, it is necessary:

- ➤ Conduct rigorous environmental and social impact studies before developing geothermal projects.

- ➤ Apply efficient and clean technologies for the extraction, treatment and re-injection of geothermal water.

- ➤ Implement emission control measures to reduce the release of gases into the atmosphere.

- ➤ Carefully monitor and manage water use to ensure availability of the resource for other uses.

- ➤ Establish transparent dialogue and participation processes with local communities to address their concerns and ensure a fair distribution of benefits.

- ➤ Comply with environmental and social standards and regulations to ensure sustainable development of geothermal energy.

Geothermal energy offers a renewable energy source with significant benefits, but it is crucial to consider and mitigate its potential environmental and social impacts to ensure sustainable development. Implementation of good practices, active participation of communities and compliance with environmental regulations are essential for geothermal energy to contribute to a clean and sustainable energy future.

3.3. Energy production technologies in today's world: A detailed analysis

Today's world faces a complex and challenging energy landscape. Growing energy demand, the depletion of traditional fossil resources, climate change and geopolitical crises have highlighted the need for a more sustainable and efficient approach to energy production. In this context, various technologies are emerging as viable alternatives to meet the energy needs of the present and the future.

3.4. Countries with the highest energy production.

Countries with the highest energy production in the world today (2024)

The global energy landscape is constantly changing, with new players emerging and technologies evolving rapidly. However, some countries remain at the forefront of energy production, meeting domestic demand and exporting surpluses to other nations. Below is an analysis of the top 10 countries with the highest energy production in the world today (2024), considering both renewable sources and fossil fuels:

1. China:

Total production: 14,100 TWh (terawatt-hours)

Main sources: Coal (62%), Hydroelectric (17%), Nuclear (5%)

Characteristics: China is the world's largest energy producer and consumer, with an energy matrix highly dependent on coal. However, it is investing heavily in renewable energies and nuclear energy to diversify its energy mix and reduce greenhouse gas emissions.

2. United States:

Total production: 10,000 TWh

Main sources: Petroleum (31%), Natural gas (30%), Coal (22%)

Characteristics: The United States is the second largest energy producer in the world, with an energy matrix based primarily on fossil fuels. However, it is also experiencing significant growth in renewable energy production, especially solar and wind energy.

3. India:

Total production: 7,100 TWh

Main sources: Coal (70%), Hydroelectric (12%), Renewable (9%)

Characteristics: India is the world's third largest energy producer and the world's largest coal consumer. Despite its dependence on coal, India is investing heavily in renewable energy, especially solar and wind power, to meet its growing energy demand.

4. Russia:

Total production: 6,700 TWh

Main sources: Petroleum (36%), Natural gas (47%), Coal (13%)

Characteristics: Russia is the world's fourth largest energy producer and a major exporter of oil and natural gas. Its energy matrix is mainly based on fossil fuels, but it is also developing its nuclear energy capacity.

5. Japan:

Total production: 3,300 TWh

Main sources: Liquefied natural gas (LNG) (40%), Coal (27%), Nuclear (21 %)

Characteristics: Japan is the world's fifth largest energy producer, but is heavily dependent on fossil fuel imports. It is implementing measures to increase energy efficiency and develop renewable energy sources, such as solar and wind power.

6. Canada:

Total production: 3,200 TWh

Main sources: Petroleum (38%), Natural gas (30%), Hydroelectric (27%)

Characteristics: Canada is the sixth largest energy producer in the world and a major exporter of oil and natural gas. Its energy matrix includes a significant proportion of hydroelectric power.

7. Saudi Arabia:

Total production: 3,100 TWh

Main sources: Petroleum (86%), Natural gas (12%)

Characteristics: Saudi Arabia is the seventh largest energy producer in the world and the largest oil exporter. Its economy is heavily dependent on oil production and exports, which poses challenges for diversifying its energy matrix and reducing its dependence on fossil fuels.

8. Germany:

Total production: 3,000 TWh

Main sources: Coal (38%), Renewables (32%), Natural gas (18%)

Characteristics: Germany is the eighth largest energy producer in the world and a leader in the transition to renewable energy. It has invested heavily in solar, wind and biomass, and aims to achieve a 100% renewable energy matrix by 2045.

9. South Korea:

Total production: 2,800 TWh

Main sources: Coal (31%), Natural gas (30%), Nuclear (26%)

Characteristics: South Korea is the ninth largest energy producer in the world and is highly dependent on fossil fuel imports. It is developing its nuclear and renewable energy capacity to reduce its dependence on imports and improve energy security.

3.5.   Global energy crisis and its implications.

The world is facing an unprecedented energy crisis, characterized by increasing demand, scarcity of fossil resources, rising prices and geopolitical repercussions. This crisis has a significant impact on all continents, affecting not only the economy and development, but also people's daily lives. The following is a detailed analysis of the global energy crisis and its implications by continent:

Europe:

Key challenges: Europe is heavily dependent on natural gas imports from Russia, which has made it vulnerable to supply disruptions and price increases following the invasion of Ukraine. The region also faces challenges in increasing renewable energy production and reducing dependence on fossil fuels.

Implications: The energy crisis in Europe has led to a significant increase in energy prices, which has generated inflation, affected business competitiveness and increased energy poverty. Governments have implemented emergency measures, such as energy rationing and subsidies, to mitigate the impact of the crisis.

North America:

Key challenges: While North America is a major producer of oil and natural gas, the region also faces challenges in increasing energy efficiency and reducing greenhouse gas emissions. In addition, the region's reliance on gasoline-based automotive transportation makes it vulnerable to oil price increases.

Implications: The energy crisis in North America has led to higher gasoline and electricity prices, impacting household and business budgets. Governments have implemented measures to increase domestic energy production and promote the use of electric vehicles.

Asia:

Main challenges: Asia is the region with the highest energy demand in the world, and its dependence on fossil fuels is high. This creates challenges for energy security, air pollution and climate change. The region also faces challenges in developing and financing large-scale renewable energy projects.

Implications: The energy crisis in Asia has led to higher energy prices, which has impacted economic growth and energy poverty. Governments in the region are implementing policies to increase energy efficiency, diversify the energy matrix and invest in renewable energy.

South America:

Main challenges: South America has great potential for renewable energy generation, but lack of infrastructure, underinvestment and political instability hinder its development. The region also faces challenges in reducing dependence on fossil fuels and improving energy efficiency.

Implications: The energy crisis in South America has led to higher energy prices, especially in import-dependent countries. Governments in the region are implementing policies to promote the development of renewable energy, improve energy efficiency and strengthen regional energy integration.

Africa:

Key challenges: Africa has great potential for renewable energy generation, but access to energy remains a major problem for many countries. Lack of infrastructure, underinvestment and poverty are additional challenges for the development of the energy sector.

Implications: The energy crisis in Africa limits economic development and increases energy poverty. Governments in the region are implementing policies to increase energy access, promote renewable energy development and strengthen regional energy cooperation.

The global energy crisis is a complex challenge with implications spanning all continents. While each region faces specific challenges, there are also opportunities for international collaboration and the development of sustainable solutions. The transition to a cleaner and more sustainable energy matrix is crucial to address the energy crisis, mitigate climate change and ensure a secure and prosperous energy future for all.

3.6. Endangered energy resources.

The growing demand for energy, dependence on fossil fuels and unsustainable extraction practices threaten the availability of various energy resources around the world. Below is an analysis of endangered energy resources by continent:

Europe:

Coal: The European Union has set ambitious targets to reduce dependence on coal, which has led to mine closures and reduced production. However, some countries still rely heavily on coal for electricity generation, raising concerns about its environmental and social impact.

Oil and natural gas: Oil and natural gas production in Europe is in decline, forcing the region to become increasingly dependent on imports. This makes it vulnerable to price increases and supply disruptions, such as those that have occurred recently due to Russia's invasion of Ukraine.

North America:

Oil: Conventional oil production in the United States is peaking, leading to a search for alternative sources such as shale oil and tar sands. However, these forms of extraction are highly polluting and water-intensive, leading to environmental and social concerns.

Natural gas: Natural gas production in the United States has increased significantly in recent years, thanks to the development of the hydraulic fracturing (fracking) technique. However, fracking has also been associated with water pollution, earthquakes and greenhouse gas emissions.

Asia:

Coal: Asia is the world's largest consumer of coal, and demand for coal continues to grow. This raises concerns about air pollution, greenhouse gas emissions and the impact on public health.

Oil and natural gas: Asia is also a major consumer of oil and natural gas, and its dependence on imports is increasing. This makes it vulnerable to price increases and supply disruptions.

South America:

Oil: Venezuela has the world's largest oil reserves, but production has declined significantly in recent years due to political and economic instability. This has raised concerns about the region's energy security.

Natural gas: Bolivia has significant natural gas reserves, but faces challenges in developing and exporting its gas due to lack of infrastructure and logistical difficulties.

Africa:

Oil: Nigeria is Africa's largest oil producer, but its oil industry faces challenges such as corruption, political instability and environmental degradation.

Natural gas: Mozambique has the largest natural gas reserves in Africa, but its development faces challenges such as lack of infrastructure, financing and security.

3.7. Economic implications of technology in energy production
Technology plays a crucial role in the production of energy, and its

evolution

has profound economic implications. A detailed analysis of the economic implications of the technology in energy production is presented below:

1. Production costs:

- Cost reduction: New technologies can reduce energy production costs in a number of ways, such as:

- Efficiency improvements: Technologies such as more efficient wind turbines, higher efficiency solar panels and higher capacity batteries reduce the amount of resources needed to generate the same amount of energy.

- Automation: Power plant process automation, such as remote turbine control and smart grid management, can reduce labor costs and improve operational efficiency.

- New energy sources: The development of renewable energy sources such as solar, wind and hydroelectric power can reduce dependence on expensive and volatile fossil fuels.

- Increased costs: However, some new technologies may also carry higher initial costs, such as:

- Research and development: Investment in research and development of new energy technologies can be significant, increasing initial implementation costs.

- Infrastructure: The construction of new infrastructure for the use of renewable energies, such as smart grids and charging stations for electric vehicles, may require significant investments.

2. Employment:

➢ Job creation: The renewable energy industry is a fast-growing sector that generates new jobs in areas such as manufacturing, installation, maintenance and research.

➢ Examples: Solar panel installation creates jobs in panel manufacturing, installation and maintenance of solar systems. Wind farm construction creates jobs in wind turbine manufacturing, wind farm construction, and facility maintenance.

➢ Job losses: The transition to renewable energy may lead to job losses in the fossil fuel industry, such as coal mining, oil and natural gas extraction, and the operation of fossil fuel power plants.

➢ Examples: Reduced coal use may lead to closure of coal mines and loss of jobs for miners. Declining demand for oil and natural gas may affect the oil and gas industry, which may lead to layoffs in exploration, production and refining companies.

➢ Recycling and reuse: The renewable energy industry can also generate jobs in the recycling and reuse of materials used in energy technologies, such as batteries and solar panels.

3. Competition:

➢ New opportunities: Technological innovation opens up new opportunities for companies and countries to compete in the global energy market.

➢ Examples: Countries that lead in the development of renewable energy technologies can become exporters of these technologies and associated services. Companies that develop innovative technologies for energy production can gain a competitive advantage in the marketplace.

➢ Intensified competition: Competition in the energy market may intensify as new technologies reduce costs and increase efficiency. This can lead to lower energy prices and greater access to energy for consumers.

4. Investment:

➢ Investment attraction: Innovative energy technologies can attract private and public capital investment, which can drive economic growth and job creation.

➢ Examples: Governments can offer tax incentives and subsidies to promote investment in renewable energy. Private companies can invest in research and development of new energy technologies with the potential for high returns on investment.

➢ Investment risks: However, investments in new energy technologies may also entail risks, such as:

➢ Technological uncertainty: The success of new energy technologies is not guaranteed, and some investments may fail.

➢ Changes in government policies: Changes in government policies that support renewable energies may negatively affect the profitability of investments in this sector.

3.8.   Energy security:

Reducing import dependence: The development of domestic renewable energy sources can reduce dependence on fossil fuel imports, which improves a country's energy security.

Example: A country that invests in solar energy can reduce its dependence on oil imports, which makes it less vulnerable to oil price fluctuations on the global market.

Energy security challenges in Panama:

1. High dependence on oil imports: Panama is highly dependent on oil imports to meet its energy demand, which makes it vulnerable to oil price fluctuations on the international market.

2. Limited renewable energy production: Despite its potential for renewable energy generation, Panama has not yet fully developed these energy sources, such as solar, wind, and hydroelectric power.

3. Deficient energy infrastructure: Panama's energy infrastructure, such as transmission and distribution networks, is in need of improvement to

   ensure a reliable and efficient energy supply.

4. Growth in energy demand: Panama's energy demand is growing due to population growth and economic development, which puts pressure on the current energy system.

Opportunities to improve energy security in Panama:

1. Develop renewable energy sources: Panama has great potential for solar, wind and hydroelectric power generation, which can reduce its dependence on oil imports and increase energy security.

2. Improving energy efficiency: Implementing energy efficiency measures in all sectors, such as the use of efficient appliances and vehicles, can reduce energy demand and improve energy security.

3. Strengthen energy infrastructure: Investing in the modernization and expansion of energy transmission and distribution networks can ensure a reliable and efficient energy supply throughout the country.

4. Promote regional energy integration: Collaboration with neighboring countries in the development of regional energy projects, such as cross-border power grids, can improve energy security and the stability of energy supply.

Strategies to achieve a secure and sustainable energy future in Panama:

1. Establish a sound regulatory framework: Implement a regulatory framework that encourages investment in renewable energy, energy efficiency and energy efficiency.

   modernization of the energy infrastructure.

2. Promote research and development: Invest in research and development of innovative energy technologies to improve efficiency and reduce the cost of renewable energy.

3. Promote education and awareness: Educate the public on the importance of energy security and responsible energy consumption

practices.

4. Encourage private sector participation: Create an enabling environment for the private sector to invest in renewable energy, energy efficiency and energy infrastructure modernization projects.

Energy security is a fundamental issue for Panama's sustainable development. By addressing existing challenges, seizing opportunities and implementing effective strategies, Panama can achieve a secure, reliable and sustainable energy future that will drive its economic growth and the well-being of its people.

3.9.   Political implications

Technology plays a crucial role in energy production, and its evolution has profound political implications at various levels, from local politics to global geopolitics. The following is a detailed analysis of the political implications of technology in energy production:

1. National policy implications:

> Energy security and energy dependence: Technology can influence a country's energy security and its dependence on external energy sources. The development of domestic renewable energy sources, such as solar and wind power, can reduce dependence on fossil fuel imports, which can have significant foreign policy and national security implications.

➢ Energy transition and climate change: Technology plays a key role in the transition to a more sustainable, low-carbon energy system. Governments implement policies to promote the development and adoption of renewable energy technologies, set energy efficiency standards and reduce greenhouse gas emissions. These policies can generate debate and controversy among different political actors and interest groups.

➢ Energy prices and affordability: Technology can affect energy prices and the affordability of energy for consumers. Advances in renewable energy technologies can reduce energy production costs, while government policies, such as subsidies and incentives, can influence the final price of energy for households and businesses. Energy affordability is an important political issue, as it can affect social welfare and economic competitiveness.

➢ Employment and social impact: The transition to a new technology-driven energy system can have a significant impact on employment and society in general. The creation of new jobs in the renewable energy sector can offset job losses in the fossil fuel industry. However, it is important that governments implement support and training policies to ensure a fair and equitable transition for affected workers.

2. Implications for international policy:

➢ International cooperation and energy agreements: Technology can facilitate international energy cooperation, driving knowledge sharing, collaboration on renewable energy projects, and the development of global standards and regulations. International energy agreements can be crucial in addressing common challenges such as climate change and energy security.

➢ Geopolitics and competition for resources: The possession and control of key energy technologies can become an important factor in global geopolitics. Countries may compete for access to energy resources and the development of advanced energy technologies, which can lead to international tensions and conflicts.

➢ International trade and intellectual property: Technology transfer, intellectual property protection and access to international markets are important aspects of international energy policy. Governments should establish appropriate regulatory frameworks to facilitate international trade in energy technologies, protect intellectual property and ensure equitable access to technologies for all countries.

3. Governance and citizen participation:

➢ Role of government and public policy: Governments play a crucial role in shaping the energy future through public policies, regulations, incentives and subsidies. Transparency, accountability and citizen participation are essential to ensure that energy policies are developed and implemented in a fair, efficient and sustainable manner.

➢ Community empowerment and local participation: Technology can empower local communities to participate in the production and management of their own energy. The development of small-scale renewable energy projects and the implementation of micro-grid systems can increase local energy autonomy and foster sustainable development at the community level.

- ➢ Challenges and opportunities for democracy: The energy transition and the implementation of new energy technologies present challenges and opportunities for democratic systems. It is important to ensure that decision-making processes are inclusive, transparent and accountable to citizens, and that the benefits of the energy transition are shared equitably among all sectors of society.

Technology has a profound impact on energy production, with policy implications ranging from local politics to global geopolitics. Governments, businesses, civil society and citizens must work together to seize the opportunities presented by technology to build a more sustainable, secure and prosperous energy future for all.

4. Geopolitical implications:

- ➢ Energy security: Technology can improve countries' energy security by reducing their dependence on fossil fuel imports and increasing the diversity of their energy sources.
- ➢ Examples: Domestic renewable energy development can reduce dependence on oil and natural gas imports, making countries less vulnerable to supply disruptions and price fluctuations in the global market.
- ➢ International cooperation: Technology can boost international energy cooperation by fostering knowledge sharing, collaboration on renewable energy projects, and the development of global standards and regulations.
- ➢ Examples: International agreements such as the Paris Agreement on climate change.

3.10. Panama's energy situation

Panama, like many countries in the world, faces challenges and opportunities in its energy sector. Although it has a relatively stable electricity system and access to various energy sources, it also faces challenges in terms of energy security, import dependence, renewable energy development and energy efficiency. The following is a comprehensive analysis of Panama's current energy situation:

1. Energy demand and supply:

   ➤ Demand: Energy demand in Panama has experienced steady growth in recent years, driven by population growth, economic development and urbanization. In 2022, electricity demand reached 8,169 GWh, with an average annual growth of 3.5% over the last decade.

   ➤ Supply: Panama's energy matrix is based mainly on hydroelectric (55%), oil-based thermal (38%) and wind (7%) sources. In 2022, hydroelectric power generation accounted for 56% of total production, followed by thermal (42%) and wind (2%).

2. Energy security:

   ➤ Dependence on imports: Panama relies heavily on imports of oil and petroleum derivatives for its thermal power generation. In 2022, oil imports accounted for 38% of total energy supply. This dependence on imported fossil fuels exposes the country to international price volatility and geopolitical risks.

   ➤ Diversification of the matrix: Panama has taken steps to diversify its energy matrix through the development of renewable energy projects, mainly solar and wind. However, progress in this field is still slow and dependence on fossil fuels remains high.

3. Challenges and opportunities:

Challenges:

> High dependence on fossil fuel imports.

> Limited development of renewable energies.

> Need to modernize the energy infrastructure.

> Energy losses in the transmission and distribution network.

> Low energy efficiency in various sectors.

Opportunities:

> High potential for solar and wind energy development.

> Private sector interest in investing in renewable energies.

> Favorable legal framework for the promotion of renewable energies.

> Opportunity to improve energy efficiency in all sectors.

> Potential to export surplus renewable energy to neighboring countries.

4. Strategies for a sustainable energy future:

> Increase renewable energy generation: Encourage the development of small and large-scale solar, wind and hydroelectric projects.

> Improve energy efficiency: Implement energy efficiency measures in buildings, industry, transportation and domestic consumption.

> Modernize energy infrastructure: Invest in the modernization of transmission and distribution networks to improve system reliability and efficiency.

> Promote research and development: Support research and

development of innovative energy technologies to reduce costs and improve energy efficiency. efficiency.

➢ Strengthen the regulatory framework: Establish a clear and transparent regulatory framework that encourages investment in renewable energy and energy efficiency.

➢ Promote education and awareness: Educate the population on the importance of rational energy use and the adoption of sustainable practices.

Panama has a challenging but promising path towards a sustainable energy future. The diversification of the energy matrix, the development of renewable energies, the improvement of energy efficiency and the modernization of infrastructure are key to achieving energy security, reducing dependence on imports and mitigating environmental impact. Collaboration between government, the private sector, academia and civil society is essential to design and implement effective strategies to ensure reliable, affordable and sustainable energy access for present and future generations.

3.11. Global demand for energy resources

The demand for energy resources is a crucial topic of analysis, as it has significant implications for the global economy, environment, geopolitics and sustainable development. The following is a comprehensive analysis of energy resource demand, both globally and by region:

1. Global energy demand:

Trends: Global energy demand has grown steadily in recent decades and is projected to continue to increase in the future, driven by population growth, economic development, and urbanization.

Factors: The main factors driving energy demand are:

➢ Population growth: The increase in the world's population generates a greater demand for energy to meet people's basic needs, such as lighting, heating, cooling and cooking.

➢ Economic development: The economic growth of countries, especially in emerging economies, leads to higher energy consumption in sectors such as industry, transportation and services.

➢ Urbanization: The increase in urban population increases energy demand for the operation of cities, including public transportation, urban lighting, and residential and commercial buildings.

2. Demand for energy resources by region:

➢ Asia: The Asia region is the world's largest energy consumer, accounting for more than 50% of global demand in 2021. Economic growth and industrialization in countries such as China and India are the main drivers of energy demand in the region.

➢ Europe: Europe is the region with the second highest energy demand, accounting for around 25% of global consumption in 2021. Demand in Europe is mainly driven by energy use in sectors such as industry, transportation and home heating.

➢ North America: North America accounts for approximately 20% of global energy demand. Consumption in this region is characterized by high energy use in sectors such as transportation, industry and electricity generation.

- ➢ South America: Energy demand in South America has grown significantly in recent years, driven by the economic development of countries such as Brazil and Argentina. However, the region still accounts for a small portion of global energy consumption.

- ➢ Africa: Energy demand in Africa is relatively low compared to other regions, but is expected to grow rapidly in the coming decades due to population growth and economic development.

3. Implications of the demand for energy resources:

- ➢ Energy security: High energy demand and dependence on imported energy resources can affect countries' energy security, making them vulnerable to price fluctuations and supply disruptions.

- ➢ Climate change: Energy production and consumption, especially from fossil fuels, are the main emitters of greenhouse gases that contribute to climate change.

- ➢ Environmental impacts: The extraction, processing and use of energy resources can have negative environmental impacts, such as air and water pollution, deforestation and soil degradation.

- ➢ Social challenges: Lack of access to affordable and reliable energy can have a negative impact on social and economic development, especially in the poorest and most marginalized communities.

4. Strategies to address energy demand:

- ➢ Energy transition: Reduce dependence on fossil fuels and increase the adoption of renewable energy sources such as solar, wind, hydroelectric and geothermal.

➢ Improving energy efficiency: Implement measures to reduce energy consumption in all sectors, including industry, transportation, buildings and households.

➢ Technological innovation: Develop new, more efficient, cleaner and more affordable energy technologies.

➢ International cooperation: Strengthen international cooperation to address global energy challenges, such as climate change and energy security.

3.12. Increase in energy costs and their impact on the family and national economy.

Rising energy costs in Panama have a significant impact on both the household and national economies. This impact is discussed in detail below:

1. Impact on the family economy:

➢ Reduced purchasing power: Families with lower incomes are the most affected by the rise in energy prices, as they spend a greater proportion of their budget on this item. This may force them to reduce spending on other essentials such as food, transportation or health care.

➢ Increased inequality: The impact of rising energy costs is more pronounced on low-income households, which may exacerbate economic inequality in the country.

➢ Stress on household budgets: Rising energy bills can create financial stress for families, making it difficult to save and plan financially.

2. Impact on the national economy:

> Rising inflation: Energy prices are an important component of the consumer price index (CPI), so their increase contributes to higher overall inflation. This can negatively affect the purchasing power of the population and the competitiveness of companies.

> Slowdown in economic growth: Rising energy costs can negatively affect the profitability of companies, which can lead to reduced investment, production and employment.

> Increased dependence on imports: Panama is highly dependent on imports of fossil fuels for its energy generation. The increase in international prices of these fuels may have a negative impact on the country's trade balance.

3. Measures to mitigate the impact:

> Subsidies and assistance programs: The government may implement subsidy or financial assistance programs to help low-income families cope with rising energy costs.

> Promote energy efficiency: Encourage efficient energy use through awareness campaigns, tax incentives and programs to support the adoption of efficient technologies.

> Development of renewable energy sources: Reduce dependence on imported fossil fuels by developing domestic renewable energy sources such as solar, wind, and hydroelectric.

> Diversification of the energy matrix: Seek new sources of energy supply, such as liquefied natural gas (LNG), to reduce dependence on a single supplier.

Rising energy costs are a challenge that affects both families and Panama's national economy. It is necessary to implement comprehensive measures to mitigate this impact, including subsidies, assistance programs, promotion of energy efficiency, development of renewable energies and diversification of the energy matrix. A proactive approach and a coordinated strategy between the government, the private sector and civil society are essential to ensure affordable, reliable and sustainable energy access for all Panamanians.

CHAPTER 4 The primary sector of the economy:

The primary sector of the economy plays a fundamental role in the economic and social development of countries. Agriculture, livestock, forestry and fisheries are essential activities for the production of food, raw materials and other goods and services. However, the primary sector also faces major challenges such as climate change, resource scarcity and environmental degradation. Sustainable practices and appropriate public policies are needed to ensure the viability of the primary sector and its contribution to the well-being of present and future generations.

4.1 Important characteristics of the primary sector of the economy:

Dependence on natural resources: The primary sector is based on the extraction and exploitation of natural resources directly from the environment, such as land, water, minerals and living organisms.

Low transformation: Primary sector products generally undergo low transformation before reaching the final consumer. Raw materials are extracted, harvested or captured in their natural state, with minimal processing.

Seasonal activities: Many primary sector activities, such as agriculture, fishing and forestry, are subject to natural cycles and seasons. This can affect production, prices and product availability.

Manual labor: The primary sector usually requires a high proportion of manual labor, especially in tasks such as agriculture, construction and mining.

Rural location: Primary sector activities are mainly concentrated in rural areas, where there are natural resources and adequate spaces for their development.

4.2 Agriculture:

Agriculture is the economic activity of growing plants and raising animals for the production of food, fiber and other products.

Types of agriculture: There are several types of agriculture, including family farming, commercial agriculture, industrial agriculture and organic agriculture.

Importance: Agriculture is fundamental to food security,
providing food for the world's population. It also generates employment, income and development in rural areas.

Challenges: Agriculture faces challenges such as climate change, water scarcity, soil degradation, biodiversity loss and competition for land.

Current situation, problems and items by province in Panama's agriculture: A detailed analysis

1. Current situation of agriculture in Panama:

Contribution to GDP: Agriculture accounts for about 2% of Panama's Gross Domestic Product (GDP) and employs about 15% of the active population.

Production: Panama's main agricultural products are: bananas, pineapple, rice, coffee, sugar cane, beef cattle, milk, cassava, corn, plantains and beans.

Exports: Agricultural exports represent an important part of the Panamanian economy, generating around US$1 billion annually. The main destinations for agricultural exports are the United States, Europe and Asia.

Consumption: Domestic consumption of agricultural products in Panama is also significant, especially for fresh produce such as fruits, vegetables and meat.

2. Problems facing agriculture in Panama:

Low productivity: Agricultural productivity in Panama is relatively low compared to other countries in the region. This is due to several factors, such as inefficient use of land and water, lack of access to modern technology, and low investment in research and development.

Climate change: Climate change is a major threat to agriculture in Panama because it can affect water availability, the frequency of extreme weather events and crop production.

Diseases and pests: Diseases and pests may affect
significantly the agricultural production, causing economic losses and jeopardizing food security.

Lack of access to markets: Small farmers often have difficulty accessing profitable markets, which limits their income and development opportunities.

Land fragmentation: Land fragmentation is a common problem in Panamanian agriculture, which hinders the implementation of efficient agricultural practices and the adoption of new technologies.

3. Items by province:

Bocas del Toro: Banana, banana, cocoa, coconut, pineapple, rice, cassava, African palm, cattle and pigs.

Colon: Bananas, African palm, pineapple, rice, cocoa, coconut, cassava, cattle, pigs and sheep.

Chiriqui: Coffee, bananas, pineapple, rice, onions, potatoes, carrots, broccoli, cauliflower, cattle, dairy and swine.

Veraguas: Cattle, dairy and swine, rice, corn, beans, cassava, African palm, coffee, cocoa.

Herrera: Cattle, dairy and swine, rice, corn, beans, cassava, sugar cane, African palm.

Los Santos: Cattle, dairy and swine, rice, corn, beans, cassava, sugar cane.

Coclé: Cattle, dairy and swine, rice, corn, beans, cassava, sugar cane, African palm.

West Panama: Cattle, dairy and swine, rice, corn, beans, cassava, vegetables, fruits.

East Panama: Cattle, dairy and swine, rice, corn, beans, cassava, vegetables, fruits.

Darien: Cassava, banana, rice, corn, beans, cocoa, coconut, cattle and pigs.

Emberá-Wounaan: Yucca, plantain, rice, corn, beans, cocoa, coconut, fishing.

Guna Yala: Coconut, banana, rice, yucca, fishing.

4. Strategies to improve agriculture in Panama:

Increase investment in research and development: Investment in research and development is needed to improve agricultural productivity, develop new crop varieties that are resistant to diseases and pests, and adopt sustainable agricultural practices.

Improving access to technology: Facilitating farmers' access to modern technology, such as efficient irrigation systems, agricultural machinery and digital tools, can help improve the sector's productivity and competitiveness.

Strengthen infrastructure: Improving rural infrastructure, including roads, ports and storage systems, can facilitate the transport of agricultural products to markets and reduce post-harvest losses.

Expanding access to credit: Providing small farmers with access to credit at competitive interest rates can enable them to invest in the modernization of their farms.

4.3 Livestock:

Livestock farming is the economic activity of raising animals for the production of meat, milk, eggs, wool, hides and skins and other products.

Types of livestock farming: There are several types of livestock farming, including extensive livestock farming, intensive livestock farming, family livestock farming and industrial livestock farming.

Importance: Livestock provides protein and other essential nutrients for human consumption. It also generates employment, income and development in rural areas.

Challenges: Livestock farming faces challenges such as environmental impact (greenhouse gas emissions, deforestation), animal welfare and competition for resources such as water and land.

Current situation of livestock farming in Panama:

Contribution to GDP: Livestock represents about 1.5% of Panama's Gross Domestic Product (GDP) and employs about 5% of the working population.

Production: Panama's main livestock products are beef, milk, eggs, chicken and pork.

Consumption: Domestic consumption of livestock products in Panama is significant, especially beef, chicken and eggs.

Exports: Exports of livestock products represent a small part of the Panamanian economy, mainly beef and dairy products. The main destinations for livestock exports are Central America and the Caribbean.

2. Problems facing livestock farming in Panama:

Low productivity: Livestock productivity in Panama is relatively low compared to other countries in the region. This is due to several factors, such as inefficient use of land and water, lack of access to modern technology, low investment in research and development, and the prevalence of diseases and pests. Climate change: Climate change is a major threat to livestock farming in Panama, as it can affect water availability, the frequency of extreme weather events and forage production.

Cattle rustling: Cattle rustling, or cattle rustling, is a serious problem affecting Panamanian livestock, causing economic losses and jeopardizing the security of the rural sector.

Lack of access to markets: Smallholders often have difficulty accessing profitable markets, which limits their income and development opportunities.

High input costs: The high cost of inputs, such as cattle feed, veterinary drugs and labor, can affect the profitability of livestock farming.

3. Main livestock areas:

Bocas del Toro: Cattle, dairy and swine.

Colón: Cattle, dairy and swine.

Chiriqui: Cattle, dairy and swine.

Veraguas: Cattle, dairy and swine.

Herrera: Cattle, dairy and swine.

Los Santos: Cattle, dairy and swine.

Coclé: Cattle, dairy and swine.

West Panama: Cattle, dairy and swine.

East Panama: Cattle, dairy and swine.

Darien: Cattle, dairy and swine.

Emberá-Wounaan: Cattle, dairy and swine.

Guna Yala: Cattle, dairy and swine.

4. Types of milk:

Whole milk: It is the milk that retains all its natural fat.

Semi-skimmed milk: Contains 50% less fat than whole milk.

Skim milk: Contains less than 1% fat.

UHT milk (Ultra High Temperature): Milk that has undergone an ultra-high heating process to eliminate bacteria and extend its shelf life.

Pasteurized milk: Milk that has been heated to a specific temperature to eliminate harmful bacteria, but retains its nutrients.

Organic milk: Milk produced from cows raised without the use of antibiotics, growth hormones or other chemicals.

5. Quackery:

Cattle rustling, or cattle rustling, is a serious problem affecting Panamanian cattle ranching, generating significant economic losses for farmers and jeopardizing the security of the rural sector. Panamanian authorities are taking measures to combat rustling, including increased police surveillance, the implementation of prevention programs, and stiffer penalties for rustlers.

6. Strategies to improve livestock farming in Panama:

Increase investment in research and development: Investment in research and development is needed to improve livestock productivity, develop new breeds of livestock that are more resistant to diseases and pests, and adopt sustainable livestock practices.

Improving access to technology: Facilitating farmers' access to modern technology, such as efficient irrigation systems, agricultural machinery and digital tools, can help improve the sector's productivity and competitiveness.

4.4 Forestry:

Forestry is the economic activity of managing and exploiting forests for the production of timber, fuelwood, paper, non-timber forest products and other ecosystem services.

Types of forestry: There are several types of forestry, including plantation forestry, selection forestry and community forestry.

Importance: Forestry provides essential resources such as timber, regulates the climate, protects biodiversity and prevents soil erosion.

Challenges: Forestry faces challenges such as illegal deforestation, unsustainable logging, forest fires and climate change.

Forestry in Panama:

1. Forestry's contribution to Panama:

Economic contribution: Forestry contributes to Panama's economy by generating employment, producing timber and other forest products, and providing ecosystem services such as carbon sequestration and climate regulation.

Timber Resources: Panama has important timber resources, including species such as mahogany, cedar, laurel and espavé. The forestry industry generates about 1% of Panama's Gross Domestic Product (GDP) and employs about 5% of the working population in the rural sector.

Forests: Panama's forests cover about 60% of the national territory and are home to a great biodiversity of flora and fauna. Forests play a fundamental role in watershed protection, soil erosion prevention, and climate change mitigation.

2. Forestry areas and sectors in Panama:

Natural Forests: Panama has a large expanse of natural forests, including tropical rainforests, dry forests, and cloud forests. These forests are sustainably harvested for timber and other non-timber forest products.

Forest plantations: Forest plantations are areas of land dedicated to the cultivation of trees for timber production. The main species planted in Panama are pine, teak, eucalyptus and melina. Forest plantations help meet the demand for timber and reduce pressure on natural forests.

Agroforestry: Agroforestry is the combination of agriculture, livestock and forestry in a single system. This practice is used to diversify production, improve soil fertility and conserve natural resources.

Community forestry: Community forestry is the management of forests by local communities for their own benefit. Community forestry contributes to community empowerment and forest conservation.

3. The Achiotines Laboratory:

Los Achiotines Laboratory is a forestry scientific research and development center located in Panama. The laboratory belongs to the Instituto Nacional de Investigaciones Forestales, Agrícolas y Pecuarias (INIFAP) and its main objective is to generate scientific and technological knowledge for sustainable forest management in Panama. Located in the province of Los Santos, district of Pedasí.

4. The Achiotines Laboratory's lines of research:

Forestry: The laboratory conducts research on the forestry of plantations, natural forest forestry, agroforestry and community forestry.

Forest ecology: Studies are conducted on the ecology of tropical forests, including tree population dynamics, plant-animal interactions, and nutrient cycling.

Forest management: Research on sustainable forest management techniques for timber production, biodiversity conservation and ecosystem services.

Non-timber forest products: The uses and commercial potential of non-timber forest products, such as fruits, bark, fibers and medicinal plants, are studied.

5. Impact of Los Achiotines Laboratory:

Los Achiotines Laboratory has made important contributions to the development of forestry in Panama. The laboratory's research has improved forest management practices, increased the productivity of forest plantations and conserved forest biodiversity. The laboratory's results have been used to develop public policies and sustainable forest management programs in Panama.

Forestry plays a fundamental role in the economic, social and environmental development of Panama. Los Achiotines Laboratory is a key institution for the advancement of forestry in the country, through scientific research, technological development and knowledge transfer to local communities and decision makers.

4.5 Fishing:

Fishing is the economic activity of catching fish, crustaceans, mollusks and other aquatic organisms for human or industrial consumption.

Types of fishing: There are several types of fishing, including artisanal fishing, industrial fishing, deep-sea fishing and aquaculture.

Importance: Fishing provides an important source of protein and other nutrients for human consumption. It also generates employment and income in coastal communities.

Challenges: Fisheries face challenges such as overfishing, marine pollution, climate change and illegal fishing.

Areas of Ichthyological Richness in Panama: Products and Current Problems

Panama, as a country with two extensive coastlines (Pacific and Caribbean) and a large number of rivers and lakes, has a great ichthyological wealth, which makes it an important producer and exporter of fishery products. However, Panama's fisheries face several challenges that threaten the sustainability of the sector and marine biodiversity.

2. Areas of Ichthyological Richness:

Pacific: The Gulf of Chiriqui, the Gulf of Panama and Coiba Island are particularly rich in fishery resources, where species such as tuna, sardines, mackerel, shrimp, snapper and corvina are found.

Caribbean: Costa Abierta, Bocas del Toro and Comarca Guna Yala are areas with great marine diversity, where species such as grouper, snook, sawfish, shrimp and lobster are found.

Rivers and lakes: The Chagres, Tuira, Changuinola and San Miguel rivers, as well as Gatun Lake, are home to freshwater species such as tilapia, catfish, guapote and tarpon.

3. Fishery Products:

Fresh and frozen fish: Tuna, sardine, mackerel, snapper, sea bass, grouper, snook, tilapia, catfish, guapote and shad.

Seafood: Shrimp, lobster, crab, octopus and squid.

Processed products: Fish meal, fish oil, canned fish and fillets.

4. Current Issues:

Overfishing: Overfishing of some species, such as tuna and shrimp, is jeopardizing the sustainability of fish stocks.

Illegal, unreported and unregulated (IUU) fishing: IUU fishing is a threat to fishery resources and marine ecosystems because it is not governed by established rules and regulations.

Marine pollution: Pollution from domestic, industrial and agricultural waste is affecting water quality and fish health.

Climate change: Rising water temperatures, ocean acidification and changing ocean current patterns are negatively impacting fish populations.

5. Strategies for Sustainability:

Sustainable fisheries management: Implement fisheries management plans that establish fishing quotas, minimum catch sizes and closed seasons to protect fish stocks.

Combat IUU fishing: Strengthen maritime surveillance and control, implement vessel tracking systems and apply harsher penalties to violators.

Marine pollution reduction: Implement policies and programs to reduce the discharge of pollutants into the sea, promote wastewater treatment, and encourage sustainable agricultural practices.

Protection of marine ecosystems: Establish marine protected areas to preserve marine biodiversity and support fish reproduction.

Research and development: Invest in scientific research to better understand fish stock dynamics and develop more sustainable fishing practices.

Panama's ichthyological wealth is a valuable resource that must be used sustainably for the benefit of present and future generations. Strategies must be implemented to combat overfishing, IUU fishing, marine pollution and the effects of climate change. Collaboration between the government, the private sector, local communities and international organizations is essential to ensure the sustainability of Panama's fisheries and the conservation of marine biodiversity.

4.6. Impact of the geographical environment on primary sector activities.

The geographical environment has a significant influence on primary sector activities, since the physical and climatic characteristics of a territory determine to a large extent what type of crops, livestock or fishing can be developed. The following is an analysis of the main factors of the geographical environment that affect primary sector activities:

4.6.1. Incidence of latitude on the geographical environment and primary activities Latitude, the angular position of a place on Earth with respect to the equator, has a significant influence on the geographical environment and, consequently, on the primary activities that can be developed in a territory.

Temperature influences the distribution of crops, as each species has an optimal temperature range for growth. For example, tropical crops such as bananas and pineapples require warm climates, while temperate crops such as wheat and barley need cooler climates.

Precipitation: The amount and distribution of rainfall are critical for agriculture, as water is essential for plant growth. Regions with high rainfall are suitable for crops that require a lot of water, such as rice, while arid regions are more suitable for drought-tolerant crops, such as cactus and agave.

Sunshine: The number of hours of sunshine per day is important for photosynthesis, the process by which plants produce their own food. Crops that require a lot of sunlight, such as sunflowers and tomatoes, do best in regions with high insolation.

The following is an analysis of the main effects of latitude on the geographical environment and its relationship with primary sector activities:

4.6.2. Weather:

Distribution of climatic zones: Latitude is the main factor that determines the distribution of climatic zones on Earth. As we move away from the equator towards the poles, the incidence of the sun's rays decreases, which causes a decrease in temperature. Thus, different climatic zones are generated, such as tropical, subtropical, temperate, polar and subarctic.

Precipitation: Latitude also influences the distribution of precipitation. In general, equatorial and subtropical regions receive more precipitation due to the convergence of updrafts and moist air currents. As we move closer to the poles, precipitation decreases due to the divergence of cold, descending air currents.

Vegetation:

Biome types: The distribution of terrestrial biomes, such as tropical forests, savannas, temperate grasslands and polar deserts, is closely related to latitude. Each biome has unique characteristics of vegetation, fauna and climate, adapted to the specific environmental conditions of its latitude.

Agricultural productivity: Latitude also influences agricultural productivity. Tropical and subtropical regions generally have greater potential for agricultural production due to warm temperatures, high rainfall and soil fertility. In temperate regions, agricultural productivity may be seasonal, while in polar regions agriculture is limited due to low temperatures and short growing season.

3. Primary activities:

Agriculture: Latitude determines what types of crops can be grown in a territory. In tropical and subtropical regions, a wide variety of fruits, vegetables and cereals, such as bananas, coffee, rice and corn, can be grown. In temperate regions, the most common crops are cereals (wheat, barley, oats), vegetables and temperate fruits (apples, pears, grapes). In the polar regions, agriculture is limited and concentrated mainly on livestock and greenhouse vegetable production.

Livestock: Latitude also influences the types of livestock that can be raised. In tropical and subtropical regions, extensive cattle, sheep and goat farming is common, taking advantage of natural pastures. In temperate regions, intensive cattle, pig and poultry farming is more common, due to the availability of feed and the necessary infrastructure. In polar regions, livestock raising is mainly limited to reindeer and other animals adapted to the cold climate.

Fisheries: Latitude also affects the distribution of marine species and fishing activity. In tropical and subtropical regions, the diversity of marine species is high and fishing is an important activity, both for local consumption and export. In temperate regions, fishing is concentrated on species such as cod, herring and salmon. In the polar regions, fishing is limited due to extreme climatic conditions and low species diversity.

4.6.3. Relief:

Altitude: Altitude influences temperature and atmospheric pressure, which affects plant growth. In general, the higher the altitude, the lower the temperature and the lower the atmospheric pressure, which limits the development of some crops. Mountainous areas are suitable for crops that tolerate cold climates and high altitudes, such as coffee and tea.

Slope: The slope of the land also plays an important role in agriculture, as it makes manual labor and the use of agricultural machinery more difficult. Areas with steep slopes are more suitable for livestock or forestry, while flat areas are more conducive to agriculture.

4.7.   Soil quality

Fertility: Soil fertility is essential for plant growth, as it provides the nutrients they need to thrive. Fertile soils, rich in organic matter and nutrients, are suitable for a wide variety of crops. Infertile soils, with low nutrient content, may require additional fertilizers to be productive.

Texture: Soil texture refers to the size of soil particles (sand, silt and clay). Soil texture affects water holding capacity and aeration, which influences plant growth. Sandy soils drain water well but retain few nutrients, while clay soils retain water and nutrients well but can be prone to compaction.

pH: The pH of the soil is a measure of its acidity or alkalinity. Most crops grow best in soils with a slightly acidic or neutral pH. Very acidic or very alkaline soils may require adjustments to make them suitable for agriculture.

4.8.   Incidence of water availability

Water is a fundamental resource for life on Earth and plays a crucial role in primary sector activities such as agriculture, livestock farming and fishing. The availability of water in a territory has a significant influence on the geographical environment and, consequently, on the possibilities for the development of these activities. The main effects of water availability on the geographical environment and its relationship with primary sector activities are discussed below:

1. Influence on the geographical environment:

Ecosystem formation: Water availability is essential for the formation and maintenance of diverse ecosystems, from tropical rainforests to arid deserts. The quantity and distribution of water determine the flora and fauna that can thrive in a place, shaping the biodiversity and natural richness of the environment.

Geological processes: Water also plays an important role in geological processes, such as erosion, sedimentation and the formation of rivers and lakes. The amount of water flowing through a territory can modify the relief of the land and create unique landscapes, such as canyons, deltas and estuaries.

2. Impact on primary sector activities:

Agriculture: Water availability is essential for agriculture, as it is indispensable for crop irrigation. Water scarcity can limit agricultural production, especially in arid and semi-arid regions. On the other hand, excess water can also be problematic, as it can lead to flooding, soil salinization and crop diseases.

Livestock: Livestock also depends on the availability of water for animal watering and pasture production. In regions with water scarcity, extensive livestock farming may be unsustainable and require alternatives such as intensive livestock farming or the breeding of drought-adapted animals.

Fisheries: Freshwater and marine fisheries depend on the quality and quantity of water for the survival of aquatic species. Water pollution, overfishing and climate change can negatively affect the availability of fishery resources.

3. Strategies for water management in the primary sector:

Efficient water use: Implementing efficient irrigation practices, such as drip irrigation or the use of treated wastewater, can help reduce water consumption in agriculture and livestock.

Protection of water sources: Preserving watersheds, controlling water pollution and promoting reforestation are essential measures to ensure water quality and availability for primary sector activities.

Development of sustainable technologies: Investing in research and development of technologies that optimize water use, such as smart irrigation systems or desalination techniques, can contribute to the sustainability of the primary sector in water-scarce regions.

Water availability is a crucial factor that determines the potential of the geographical environment for primary sector activities. Proper water management, through sustainable practices and innovative technologies, is essential to ensure food production, ecosystem conservation and sustainable rural development. Collaboration between governments, communities and the private sector is essential to address the challenges related to water scarcity and promote efficient and responsible use of this vital resource.

4.9. Participation of the primary sector in the national economy.

Participation of the primary sector in the Panamanian economy: Current analysis and perspectives

The primary sector, also known as the agricultural sector, comprises economic activities related to the exploitation of natural resources, such as agriculture, livestock, forestry and fishing. This sector plays a fundamental role in Panama's economy, contributing to food security, employment generation and rural development.

1. Contribution of the primary sector to GDP:

Percentage of GDP: According to data from Panama's National Institute of Statistics and Census (INEC), in 2022, the primary sector contributed 2.3% to the national Gross Domestic Product (GDP). While this percentage may seem relatively low, it is important to consider that the primary sector has a significant impact on other sectors of the economy, such as the food industry and tourism.

Comparison with other sectors: Compared to the tertiary sector (services), which accounts for about 73% of GDP, and the secondary sector (industry), which contributes 24.7%, the primary sector's share of the Panamanian economy is lower. However, it is important to note that the primary sector plays a crucial role in providing basic foodstuffs and raw materials for other sectors.

2. Importance of the primary sector in Panama:

Food security: The primary sector guarantees the production of basic foodstuffs for the Panamanian population, contributing to food security and food sovereignty in the country. Family farming and small-scale agriculture play a fundamental role in the production of fresh and nutritious food for local consumption.

Employment generation: The primary sector is an important source of employment in Panama, especially in rural areas. According to INEC data, in 2020, the agricultural sector employed about 15.7% of the country's employed population.

Rural development: The primary sector contributes to sustainable rural development, promoting income generation, improved infrastructure and the conservation of natural resources in rural areas. Agriculture and livestock farming can be tools for combating poverty and improving the quality of life of rural communities.

3. Challenges and opportunities for the primary sector in Panama:

Challenges: The primary sector in Panama faces a number of challenges, such as low productivity, lack of access to credit and technology, international competition, climate change and environmental degradation.

Opportunities: Despite the challenges, the primary sector in Panama also presents significant opportunities. The growth in global food demand, the adoption of innovative technologies, the diversification of production and the promotion of value-added agricultural products are some of the opportunities that can be used to strengthen the sector.

4. Strategies to strengthen the primary sector in Panama:

Investment in research and development: Investment in research and development is needed to improve crop and livestock productivity, develop new crop varieties and livestock breeds that are more resistant to diseases and pests, and adopt sustainable agricultural practices.

Access to credit and financing: Facilitating small farmers' access to credit and financing at competitive rates can contribute to modernizing production, acquiring technology and improving infrastructure.

Training and technical assistance: Providing training and technical assistance to farmers on topics such as good agricultural practices, pest and disease management, marketing and business management can strengthen their capacities and improve their competitiveness.

Promotion of agroindustry: Promoting the development of agroindustry, adding value to agricultural and livestock products, can generate greater income for producers and diversify the rural economy.

Environmental protection: Implementing sustainable agricultural and livestock practices that preserve natural resources and protect the environment is essential for the long-term sustainability of the primary sector.

The primary sector plays a fundamental role in Panama's economy, contributing to food security, employment generation and rural development. Despite the challenges it faces, the sector also presents important opportunities to strengthen itself and contribute to the country's economic growth. The implementation of appropriate strategies, such as investment in research and development, access to credit and financing, training and technical assistance, promotion of agribusiness, and environmental protection, are key to fostering the sustainable development of Panama's primary sector.

4.10. Panama's employed population in the primary sector: Data and Analysis

1. Current figures:

Percentage of the employed population: According to data from Panama's National Institute of Statistics and Census (INEC), in 2022, the primary sector occupied 15.7% of the country's employed population. This represents 273,139 people engaged in agricultural, livestock, forestry and fishing activities.

Distribution by sex: Of the population employed in the primary sector, 52.29% are men and 47.71% are women. This distribution reflects the significant participation of women in family farming and small-scale agriculture, especially in rural areas.

Geographic distribution: The majority of the population employed in the primary sector is located in rural areas of the country. In the year 2022, 79.9% of the people employed in this sector resided in rural areas, while 20.1% resided in urban areas.

2. Historical evolution:

Decreasing trend: In recent decades, the share of the primary sector in Panama's employed population has shown a decreasing trend. In 1990, the primary sector occupied 26.7% of the employed population, while in 2022, this figure was reduced to 15.7%.

Factors affecting the decline: This decline is due to various factors, such as urbanization, diversification of the economy towards the tertiary sector (services), low productivity in the primary sector, and migration of the rural population to the cities in search of better job opportunities.

3. Importance of the primary sector:

Despite the decline in its share of the employed population, the primary sector continues to be an important sector in the Panamanian economy, as it is the main source of income for the country's economy:

Guarantees food security: The primary sector produces the basic foodstuffs consumed by the Panamanian population, contributing to the country's food sovereignty.

Generates employment: The primary sector is an important source of employment, especially in rural areas, where employment opportunities in other sectors are limited.

Contribute to rural development: The primary sector contributes to sustainable rural development, promoting income generation, infrastructure improvement and the conservation of natural resources in rural areas.

4. Challenges and opportunities:

Challenges: The primary sector in Panama faces a number of challenges, such as low productivity, lack of access to credit and technology, international competition, climate change and environmental degradation.

Opportunities: Despite the challenges, the primary sector also presents significant opportunities, such as the growth in global food demand, the adoption of innovative technologies, diversification of production and the promotion of value-added agricultural products.

The population employed in the primary sector in Panama has decreased in recent decades, but it continues to be an important sector for the country's economy and rural development.

It is necessary to implement strategies to strengthen the primary sector, such as investment in research and development, access to credit and financing, training and technical assistance, promotion of agroindustry and environmental protection.

The primary sector has the potential to contribute to Panama's sustainable economic growth, generating jobs, improving the quality of life of rural communities and ensuring the country's food security.

4.11. Participation of the Employed Population in Panama's GDP: Analysis and Perspectives

1. Contribution of the employed population to GDP:

Panama's employed population plays a fundamental role in the generation of the country's Gross Domestic Product (GDP). According to data from Panama's National Institute of Statistics and Census (INEC), in the year 2022, the total employed population will amount to 1,741,127 people, which represents 66.5% of the population aged 15 years and older.

2. Distribution by economic sector:

The share of the employed population in GDP is distributed as follows according to economic sectors:

Tertiary sector (services): The tertiary sector has the largest share of GDP, at 73%. This reflects the importance of services in the Panamanian economy, including activities such as commerce, tourism, finance, and professional services.

Secondary sector (industry): The secondary sector contributes 24.7% of GDP. This sector includes activities such as manufacturing, construction and mining.

Primary sector (agriculture and livestock): The primary sector has a minor share in GDP, at 2.3%. However, as mentioned above, this sector plays a crucial role in the country's food security and rural development.

3. Historical evolution:

The distribution of the employed population by economic sector has shown some changes in recent decades:

Tertiary sector: The tertiary sector's share of GDP has increased significantly, from 65.2% in 1990 to 73% in 2022. This is due to the growth of activities such as tourism, international trade and financial services.

Secondary sector: The secondary sector's share of GDP has decreased slightly, from 28.1% in 1990 to 24.7% in 2022. This is due, in part, to the country's trade liberalization and increased international competition.

Primary sector: The primary sector's share of GDP has declined more sharply, from 6.7% in 1990 to 2.3% in 2022. This decline is due to several factors, such as the sector's low productivity, lack of access to credit and technology, and the migration of the rural population to the cities.

4. Impact of the COVID-19 pandemic:

The COVID-19 pandemic had a negative impact on the Panamanian economy, especially in the tertiary sector, where the activities most affected by the social restriction measures are concentrated. The employed population in the tertiary sector decreased by 14.6% in 2020, while the employed population in the primary sector remained relatively stable.

5. Future prospects:

Panama's economy is expected to recover gradually in the coming years, driven by growth in the tertiary sector and investment in infrastructure. However, it is important to implement strategies to strengthen the primary sector and improve its productivity, so that this sector can contribute more significantly to the country's GDP and rural development.

Panama's employed population plays a fundamental role in the generation of the country's GDP and economic development. The distribution of the employed population by economic sector has changed in recent decades, with a growth in the tertiary sector and a decrease in the primary sector. The COVID-19 pandemic had a negative impact on the economy, especially in the tertiary sector. The economy is expected to recover in the coming years, but the primary sector needs to be strengthened to contribute more significantly to GDP and rural development.

4.12. Reality of agricultural, livestock, forestry and fishery production in Panama: A comprehensive analysis

1. Overview:

Agricultural, livestock, forestry and fisheries production in Panama plays a fundamental role in the country's economy and development, especially in rural areas. However, this sector faces various challenges that limit its potential and affect the quality of life of the people who depend on it.

2. Agricultural production:

Progress: Progress has been made in the production of some crops, such as bananas, pineapples, and rice, which are important export products.

Challenges: Low productivity, lack of access to credit and technology, international competition, climate change and environmental degradation are some of the main challenges facing Panamanian agriculture.

Opportunities: Diversification of production, adoption of sustainable agricultural practices, promotion of value-added agricultural products and strengthening of value chains are some of the opportunities that can be seized to improve the competitiveness of the agricultural sector.

3. Livestock production:

Progress: Cattle and swine are the main livestock activities in Panama. The genetic quality of livestock has been improved and meat and milk production has increased.

Challenges: Lack of quality pasture, low productivity, high dependence on imported inputs and international competition are some of the main challenges facing Panamanian livestock farming.

Opportunities: Technification of production, promotion of sustainable livestock farming, diversification of production into goat and sheep farming, and development of markets for value-added meat and dairy products are some of the opportunities that can be seized to strengthen the livestock sector.

4. Silvicultural production:

Progress: Reforestation programs have increased forest cover in some areas of the country.

Challenges: Illegal deforestation, indiscriminate logging, lack of sustainable forest management and competition from other land uses are some of the main challenges facing Panamanian forestry.

Opportunities: Promotion of sustainable forestry, development of commercial forest plantations, industrialization of timber and diversification of production into non-timber products are some of the opportunities that can be seized to strengthen the forestry sector.

5. Fish production:

Progress: Production of some fish species, such as tilapia and shrimp, has increased thanks to aquaculture.

Challenges: Overfishing, illegal, unreported and unregulated (IUU) fishing, water pollution and climate change are some of the main challenges facing Panamanian fisheries.

Opportunities: Promoting sustainable fisheries, developing marine aquaculture, diversifying production towards new fish species and adding value to fishery products are some of the opportunities that can be seized to strengthen the fisheries sector.

Agricultural, livestock, forestry and fisheries production in Panama has great potential to contribute to the country's economic and social development. However, it is necessary to address the challenges facing this sector so that it can reach its full potential. The implementation of appropriate strategies, such as investment in research and development, access to credit and financing, training and technical assistance, promotion of sustainable practices and environmental protection, are key to boosting the sustainable development of the primary sector in Panama.

4.13. Policies oriented to the development of primary sector activities.

4.13.1.   Government support:

The Government of Panama has implemented various policies and strategies to support the development of the primary sector in order to strengthen agriculture, livestock, forestry and fisheries and contribute to the country's economic growth, food security and rural development.

The following is an analysis of the main government initiatives in this area:

1. Public policies:

National Agricultural Development Plan (2021-2025): This plan defines the strategic guidelines for the development of the primary sector in Panama, with emphasis on the modernization of production, promotion of agribusiness, environmental sustainability, and social inclusion.

Agricultural Development Law (Law 38 of 2013): This law establishes the legal framework for the promotion of agricultural production, agribusiness and the commercialization of agricultural products.

National Food and Nutritional Security Strategy (2017-2022): This strategy seeks to guarantee access to safe and nutritious food for the entire Panamanian population, with an emphasis on local production and support for small family farming.

2. Government programs and projects:

Agro Rural Program: This program provides technical and financial assistance and training to small farmers to improve their productivity and competitiveness.

National Seed Program: This program promotes the production and use of high quality seeds to improve agricultural yields and resistance to pests and diseases.

National Agricultural Credit Program: This program facilitates access to credit for agricultural producers through subsidized interest rates and flexible payment terms.

National Agricultural Health Program: This program seeks to prevent and control pests and diseases affecting crops and livestock in order to protect national agricultural production.

3. Tax incentives:

Law for the Attraction of Investments in the Agricultural Sector (Law 69 of 2019): This law grants tax benefits to companies investing in the agricultural sector, such as income tax exemption and the importation of machinery and equipment.

Income Tax Exemption for Agricultural Production: This law exempts income generated by agricultural production from income tax, thus encouraging investment in this sector.

4. Institutional strengthening:

Modernization of the Ministry of Agricultural Development (MIDA): MIDA has implemented a modernization process to improve its efficiency and effectiveness in serving agricultural producers.

Creation of the Instituto de Innovación Agropecuaria de Panamá (IIAP): The IIAP is an institution dedicated to research and development of new agricultural technologies to boost the productivity and sustainability of the primary sector.

Strengthening producer organizations: The government has provided support to producer organizations to strengthen their capacity to manage, negotiate and market agricultural products.

5. Challenges and perspectives:

Despite the government's efforts, the primary sector in Panama still faces a number of challenges, such as low productivity, lack of access to credit and technology, international competition, climate change and environmental degradation. To overcome these challenges and take advantage of the opportunities presented by the primary sector, it is necessary to continue implementing effective public policies, strengthen agricultural sector institutions, encourage private investment, and promote the adoption of sustainable practices.

In conclusion, the Panamanian government has made a significant effort to support the development of the primary sector through various policies, programs, projects and incentives. However, much remains to be done to strengthen this sector and reach its full potential. Collaboration between the government, the private sector, academia and producer organizations is essential to boost the sustainable development of the primary sector in Panama and contribute to economic growth, food security and the welfare of rural communities.

4.13.2. Support from international organizations for production

Panama, like many developing countries, receives support from various international organizations to strengthen its primary sector, which includes agricultural, livestock, forestry and fishing activities. This support takes the form of financing, technical assistance, training, the exchange of knowledge and best practices, and the promotion of public policies favorable to the sustainable development of the sector.

1. Main international agencies providing support:

Food and Agriculture Organization of the United Nations (FAO): FAO works with the Panamanian government to improve agricultural production, food security and nutrition, natural resource management and sustainable rural development.

World Bank (WB): The WB finances agricultural, livestock, forestry and fisheries development projects and provides technical assistance to improve the productivity and competitiveness of the primary sector.

International Fund for Agricultural Development (IFAD): IFAD finances small-scale family farming projects, with the aim of improving the living conditions of rural communities and strengthening food security.

Inter-American Development Bank (IDB): The IDB finances rural infrastructure projects, technological development, marketing of agricultural products and climate risk management in the primary sector.

United States Agency for International Development (USAID): USAID provides technical and financial assistance to improve food security, nutrition, sustainable agriculture and natural resource management in Panama.

2. Types of support provided:

Financing: International agencies provide loans and grants to the Panamanian government and private sector organizations to finance agricultural, livestock, forestry, and fisheries development projects.

Technical assistance: International organizations provide technical assistance to government authorities, agricultural producers, civil society organizations and other stakeholders in the primary sector to improve their capacities in areas such as sustainable agricultural production, natural resource management, marketing of agricultural products and adaptation to climate change.

Training: International organizations offer training programs for agricultural producers, agricultural technicians, extensionists and other primary sector actors, with the objective of strengthening their knowledge and skills to improve the sector's productivity and competitiveness.

Exchange of knowledge and best practices: International organizations facilitate the exchange of knowledge and good practices between countries, so that Panama can learn from successful experiences in other countries and apply them in its own context.

Promotion of public policies: International agencies work with the Panamanian government to promote public policies that favor the sustainable development of the primary sector, such as investment in agricultural research and development, access to credit for producers, protection of natural resources, and marketing of agricultural products.

3. Impact of international support:

The support of international organizations has had a positive impact on Panama's primary sector, contributing to:

Increased agricultural productivity: Technical assistance and financing have enabled farmers to adopt new technologies and more efficient agricultural practices, resulting in increased food production.

Improved food security: Support for small-scale family farming and the promotion of sustainable agriculture have contributed to improving food security in rural communities.

Natural resource conservation: Natural resource management and climate change adaptation programs have helped protect forests, water and other natural resources important to the primary sector.

Strengthening primary sector institutions: Technical assistance and training have strengthened the capacities of government institutions and private sector organizations involved in primary sector development.

4. Challenges and perspectives:

Despite the positive impact of international support, the primary sector in Panama still faces challenges such as low productivity in some areas, lack of access to credit for small producers, international competition, climate change and environmental degradation. To overcome these challenges and take advantage of the opportunities presented by the primary sector, it is necessary to continue working in collaboration with international organizations, strengthen public policies, promote private investment and adopt sustainable practices.

The support of international organizations plays a fundamental role in the development of the primary sector in Panama. This support takes the form of financing, technical assistance, training, the exchange of knowledge and best practices, and the promotion of public policies favorable to the sustainable development of the sector. The positive impact of this support is reflected in increased agricultural productivity, improved food security, the conservation of natural resources and the strengthening of primary sector institutions.

4.13.3. Mechanisms to ensure food safety

Food security, defined as access for all people at all times to sufficient, safe and nutritious food to meet their nutritional needs and food preferences, is a fundamental issue for Panama's sustainable development. The Panamanian government has implemented various mechanisms to guarantee food security for its population, which can be grouped into the following categories:

1. Increase in domestic production:

Strengthening family agriculture: Support is provided to small farmers to improve their productivity through technical assistance, financing, training and access to markets.

Promotion of sustainable agriculture: The adoption of sustainable agricultural practices that protect the environment and reduce dependence on external inputs is encouraged.

Investment in research and development: Support is provided for research and development of new agricultural technologies to increase crop productivity and resilience.

2. Diversification of production:

Promotion of non-traditional crops: The production of non-traditional crops with high nutritional and commercial value is encouraged in order to reduce dependence on imports of staple foods.

Aquaculture development: Support is provided for the development of aquaculture as an alternative source of protein, especially in areas where agriculture is limited.

Strengthening sustainable livestock farming: Sustainable livestock farming that protects the environment and improves the quality of meat and milk is promoted.

3. Access to food:

Social assistance programs: Social assistance programs are implemented to provide food to the most vulnerable populations, such as children, pregnant women and the elderly.

Food banks: Food banks are established that collect and distribute food to people in need.

Popular markets: We encourage the creation of popular markets where consumers can buy fresh food at affordable prices.

4. Reduction of food losses and waste:

Awareness campaigns: Awareness campaigns are conducted to raise awareness of the importance of reducing food losses and waste.

Strengthening value chains: Support is provided to strengthen agricultural value chains to reduce post-harvest losses and improve the efficiency of food distribution.

Promotion of food reuse and recycling: The reuse of food surpluses for the production of new products and the recycling of organic waste for compost production are encouraged.

5. Governance and coordination:

National Council for Food and Nutritional Security (CONASAN): CONASAN is the governing body for food security in Panama, and coordinates the actions of the different government institutions and civil society in this area.

National Food and Nutritional Security Plan (PLANSAN): PLANSAN is the planning instrument for public policy on food security in Panama, and establishes the objectives, strategies and actions to guarantee access to sufficient, safe and nutritious food for the entire population.

Monitoring and evaluation: Periodic studies and analyses are carried out to evaluate the food security situation in Panama and the effectiveness of the policies implemented.

The Panamanian government has implemented various mechanisms to ensure food security for its population. These mechanisms range from increasing domestic production and diversifying production, to access to food, reducing food losses and waste, and the governance and coordination of food security actions. While significant progress has been made, much remains to be done to ensure that all people in Panama have access to sufficient, safe and nutritious food to meet their nutritional needs and food preferences. Effective implementation of public policies, collaboration between government, the private sector and civil society, and investment in research and development are key to achieving this goal.

VOCABULARY

1. Irrigated Agriculture: A type of agriculture that uses irrigation systems to provide water to crops, usually through canals, pipes or other methods.
2. Rainfed agriculture: This is a type of agriculture that relies on rainfall and natural soil moisture, without using artificial irrigation systems.
3. Subsistence Agriculture: A form of agriculture in which farmers produce food primarily for their own and their family's consumption, rather than for sale in the market.
4. Shifting agriculture: Also known as shifting or swidden agriculture, this is an agricultural system in which
5. Agrotourism: A form of tourism that combines activities related to agriculture and the rural environment. Tourists visit farms, participate in agricultural activities, learn about local products and experience rural life, thus contributing to the economic development of rural areas.
6. Airbnb: It is an online platform that allows people to rent or rent short-term accommodations, such as rooms, apartments or houses, through the intermediation of the platform. Airbnb has had a significant impact on the hospitality industry and has generated changes in the tourist rental market.
7. Fallow: An agricultural practice that consists of leaving a plot of land uncultivated for one or more periods of time, in order to allow the soil to recover and regenerate.
8. Money Laundering: Also known as money laundering, it is the process by which the illicit origin of funds or assets is concealed, passing them off as legitimate through complex and difficult to trace financial transactions, with the aim of integrating them into the legal system and avoiding detection.
9. Working Capital: Refers to the cash and liquid assets that a company uses in its daily operations, such as cash, accounts receivable and inventories.
10. Capitalism: An economic system in which the means of production and distribution are in the hands of individuals and private companies, with the objective of obtaining profits.
11. Economic Cycle: Refers to the periodic fluctuations experienced by an economy in terms of growth and contraction. It includes phases of expansion, recession, depression and recovery.

12. Cluster: A geographic concentration of companies, suppliers and other related institutions in a specific field. The cluster can foster collaboration and competition among companies and generate economic benefits.

13. Conurbation: The process of growth and merging of nearby urban areas, forming a continuous city or metropolitan area.

14. Economic Growth: The sustained increase in the production of goods and services in an economy over time, generally measured by the increase in Gross Domestic Product (GDP).

15. Economic Crisis: A situation of great instability and difficulties in an economy, characterized by a significant drop in economic growth, increase in unemployment, decrease in production and other negative indicators.

16. Degrowth: A current of thought that proposes the deliberate and planned reduction of economic growth in order to achieve a more sustainable and equitable development.

17. Decentralization: The process of transferring power and authority from a central level or government to lower levels or local entities, allowing for greater participation and decision-making at the local level.

18. Alternative Economy: Refers to economic systems and models different from traditional capitalism, such as the solidarity economy, the collaborative economy and other forms of economic organization that seek sustainability and equity.

19. Informal Economy: The sector of the economy that comprises unregulated or officially unregistered economic activities, generally carried out by self-employed workers or small businesses.

20. Local Economy: Refers to economic activity centered on a specific region or locality, where resources and consumption are primarily focused on the local community.

21. Regional Economics: The study of the economy of a specific geographic region, taking into account the economic factors and interactions specific to that region.

22. Sustainable Economy: An economic approach that seeks to meet present needs without compromising the ability of future generations to meet their own needs, taking into account environmental, social and economic aspects.

23. Franchise: A business model in which a company (franchisor) grants another company or individual (franchisee) the right to operate a business using its brand, products and systems, in exchange for royalties and compliance with certain conditions.

24. GAFAM (Google, Apple, Facebook, Amazon and Microsoft): These are the five largest and best-known technology companies in the world, which have had a significant impact on the industry and the global economy.

25. Globalization: Is the process of interconnection and growing interdependence between countries at the economic level.

26. Hinterland: The area or region behind or around an urban center or port, and which is economically influenced by that center. It refers to the zone of influence and economic activities related to the center.

27. Commercial hub refers to a center or neuralgic point in which different commercial activities converge, such as the exchange of goods, services and financial transactions. It is a place where a wide range of businesses, stores, financial institutions and related services are concentrated, creating a favorable environment for economic activity and trade.

28. Digital Footprint: It is the trace left by a person on the Internet through their online activity, including personal information, interactions in social networks, search history and other data generated on the web.

29. HDI: The Human Development Index (HDI) is an indicator that measures a country's human development, taking into account factors such as life expectancy, education and per capita income. The HDI is used to compare and rank countries according to their level of human development.

30. Economic Innovation: Refers to the introduction and application of new ideas, technologies, processes or business models in the economic sphere, with the objective of improving efficiency, productivity and competitiveness.

31. Latifundio: A large agricultural landholding owned by a single person or entity and used for large-scale agricultural production.

32. Smallholding: A small agricultural landholding owned by a farmer or family and used for small-scale agricultural production.

33. Globalization: A term used to describe the process of increasing interconnectedness and globalization of the economy, culture, politics and

other aspects of society worldwide.

34. Supply: In economics, refers to the quantity of goods or services that producers are willing to sell in the market at a given price.

35. Offshore: Refers to the location or activity carried out outside the national territory or away from the coast. It may refer to the location of companies or financial activities in jurisdictions with more favorable tax benefits or regulations.

36. Oligopoly: A market structure in which a small number of companies control the majority of the market for a given product or service,

37. WTO: The World Trade Organization. It is an international institution responsible for establishing and enforcing international trade rules among member countries. Its main objective is to promote trade liberalization and facilitate the negotiation of trade agreements.

38. OPEC: The Organization of Petroleum Exporting Countries (OPEC) is an international organization composed of oil producing countries. Its main objective is to coordinate and unify the oil policies of its members in order to stabilize oil prices in the international market.

39. Tax haven: A term used to describe a jurisdiction that offers tax benefits and lax regulations to individuals and companies, often used to avoid taxes or circumvent financial regulations.

40. GDP (Gross Domestic Product): An indicator that measures the value of all final goods and services produced in a country during a given period. It is used as a measure of the size and economic growth of a nation.

41. Pisciculture: The breeding and production of fish in ponds, lakes or other aquatic facilities, for commercial or consumption purposes.

42. Post-productivism: A current of thought that questions the traditional focus on economic growth and production, and seeks alternative forms of economic organization and sustainable development.

43. Economic Precarization: Refers to the process of increasing instability and precariousness in employment and working conditions, such as lack of job security, low wages, temporary contracts, etc.

44. Protectionism: An economic policy that seeks to protect domestic industry and a country's domestic economy through trade barriers, such as tariffs or quotas, in order to limit foreign competition and favor domestic

production.

45. SMEs (Small and Medium-Sized Enterprises): These are small or medium-sized companies, which generally have a limited number of employees and a specific annual turnover according to the criteria established in each country.

46. Economic Resources: These are the various elements and assets used in the production of goods and services, such as land, labor, capital and technology.

47. Economic Networks: Refers to the interconnections and relationships between companies, institutions and economic actors that collaborate and mutually benefit each other in a given area or sector.

48. Metropolitan Region: Refers to a geographic region that includes a major city and its surrounding areas, forming an integrated urban and economic entity.

49. Per Capita Income: The average income earned by each individual in a given area or country. It is calculated by dividing the total income of an area by its population.

50. Quaternary Sector: It is a classification of the economic sector that includes activities related to research, development, technology, information and knowledge. It is considered a knowledge-based sector of the economy.

51. Silviculture: The management and cultivation of forests for the sustainable production of timber, forest products and conservation of natural resources.

52. Union: An association of workers created to defend and promote their labor rights and working conditions. Unions bargain collectively with employers on behalf of workers and may take actions such as strikes to press for better working conditions.

53. Economic Tertiarization: Refers to the process of growth and predominance of the tertiary sector of the economy, which includes service activities such as trade, transportation, tourism, finance, among others.

54. Transnational: These are companies that operate in more than one country, having subsidiaries, branches or significant presence in different parts of the world. They are also known as multinational companies.

55. Transhumance: A form of grazing in which animals move seasonally between different grazing areas, generally in search of better weather and food conditions.

56. Global Triad: Refers to the three most important economic and political blocs in the world economy: the United States, the European Union and Japan. They account for a significant share of global production and trade.

57. Poverty Threshold: The minimum level of income or resources necessary to meet basic needs and avoid poverty. The poverty line can vary according to the context and is used to measure the incidence and severity of poverty in a given population or area.

58. Free Trade Zone: Is a delimited geographic area within a country that has special customs and tax regimes. In a free zone, benefits such as tax exemptions, simplification of customs procedures and facilities for international trade are offered in order to promote investment and economic activity.

BIBLIOGRAPHIC REFERENCES

Acemoglu, D., & Robinson, J. A. (2012). Why nations fail: The origins of power, prosperity, and poverty. Crown Publishing Group. 568 pp.

Atencio Peñaloza, M. O. (2009). Cooperative legislation in Panama.

Carrera, R. R. (1964). Land law in the agrarian reform laws of Latin America. *Revista de Estudios Agrosociales, 48,* 153.

Davis, D. R., Henderson, J. V., & O'Brien, A. J. (2020). Regional and urban economics (7th ed.). McGraw-Hill. 784 pp.

Del Rosario Guerra, C. (2007). Development of the National Agroecological Zoning Program of Panama: an Ecosystem Approach. *Application of the Ecosystem Approach in Latin America,* 69.

Duranton, G., & Puga, D. (2004). Development, disparities, and distance: A quantitative analysis. MIT Press. 552 pp.

Florida, R. (2002). The rise of the creative class: And how it's transforming work, leisure, life, and politics. HarperBusiness. 336 pp.

Fujita, M., Krugman, P. R., & Venables, A. J. (2016). The new economic geography: Effects of localization, specialization, and trade. MIT Press. 552 pp.

García, J. M. F. (1970). Agrarian counter-reform and rural cadastre in Panama. *Revista de economía política, (54),* 134.

Glaeser, E. L., & Gottlieb, B. R. (2009). The economics of cities. MIT Press. 512 pp.

Grossman, G. M., & Helpman, E. (1991). Innovation and growth in the global economy. MIT Press. 528 pp.

Helpman, E., & Krugman, P. R. (1985). Market structure, trade, and foreign investment. MIT Press. 312 pp.

Henderson, J. V., & Wang, F. (2004). The effects of urban spatial structure on economic performance. In H. J. J. Sonstelie, & P. M. Townroe (Eds.), Handbook of regional and urban economics (Vol. 4, pp. 363-418). Elsevier. 624 pp.

Kolstad, C. D., & Rezai, M. (2012). The economic effects of climate change: A review of studies. Ecological Economics, 85, 33-47.

Krugman, P. R. (1997). What makes a currency area? In S. B. Cohen, & J. A. Williamson (Eds.), A world of currency areas (pp. 1-28). Cambridge University Press. 344 pp.

Krugman, P. R. (1998). Geography and trade. MIT Press. 304 pp.

Krugman, P. R., & Obstfeld, M. (2008). International economics: Theory and policy (9th ed.). Pearson Education. 832 pp.

Krugman, P. R., & Obstfeld, M. (2023). International economics: Theory and policy (11th ed.). Pearson Education. 848 pp.

Krugman, P. R., & Venables, A. J. (1995). Spatially integrated economies: Modeling the international economy. MIT Press. 344 pp.

Lucas, R. E. (1988). Lectures on economic growth. Harvard University Press. 320 pp.

Lucas, R. E., & Weber, E. (2010). Climate change and economic growth. The American Economic Review, 100(4), 81-102.

Ortega, A. A. (2021). La Prescripción Adquisitiva de Dominio Civil y la Prescripción Adquisitiva Agraria en la Legislación Panameña. *Revista Jurídica Latinoamericana, 1*(1), 29-33.

Panama, I. N. E. C. National Institute of Statistics and Census of Panama [On line]. *INEC Panama.*

Porter, M. E. (1990). The competitive advantage of nations: Creating and sustaining superior performance. Free Press. 596 pp.

Romer, P. M. (1994). The theory of endogenous growth. Cambridge University Press. 248 pp.

Sachs, J. D. (2005). The end of poverty: Economic possibilities for our time. Penguin Books. 416 pp.

Sen, A. K. (1999). Development as freedom. Oxford University Press. 368 pp.

Stern, N. (2007). The economics of climate change: The stern review. Cambridge University Press. 712 pp.

Valverde Batista, R. A. (2021). Actions and consequences of the Panamanian economic policy in the Primary Sector: generating a proposal for an economic, social and environmental model.

Printed by Books on Demand GmbH, Norderstedt / Germany